Ham Radio General Class License Study Guide

Upgrade Your License! Move Beyond the Basics of Amateur Radio
Ace the FCC Exam and Expand Your On-Air Reach

Morse Code Publishing

Contents

Introduction

WELCOME TO THE NEXT chapter in your amateur radio journey!

I'm thrilled to be a part of it. Not long ago, I was right where you are—excited to go beyond my Technician Class License and eager to earn my General.

This upgrade opened up a new world of long-distance communications that I was excited to try out. I wrote this book to help you experience that excitement based on what I found helped me pass my exam.

This book is your study guide to that next chapter – the General Class License.

With *"Ham Radio General Class License Study Guide: Upgrade Your License! Move Beyond the Basics of Amateur Radio, Ace the FCC Exam, and Expand Your On-Air Reach,"* you're getting a roadmap designed to help you ace the FCC exam and gain access to new frequencies, modes, and connections.

This guide goes beyond simply passing the test; it is about building a richer understanding of amateur radio. I've included all the FCC exam questions, correct answers, practice exams, and practical strategies to help make the material stick. My goal is to simplify your study process, keep you engaged, and give you the confidence to tackle this exam.

Consider this book as your reliable companion on your General Class license path. It's here to make even the trickiest parts of the exam feel straightforward, preparing you not only to pass but to truly appreciate the expanded privileges that come with your new license.

So, if you're a current ham operator ready to take that next big step or someone passionate about deepening your understanding of amateur radio, this book is for you.

The book is organized into four main sections: Rules & Regulations, Operating Your Radio, The Physical Radio Station, and The Science of Radio. Each section aligns with the sub-elements of the General Class License exam. This structure ensures comprehensive coverage of all topics, making it easy to navigate and focus on specific areas.

One of the unique points of this book is its practical approach. It combines theoretical knowledge with real-world applications, making the learning process relevant and engaging. Memorable mnemonics, clear explanations, and practical examples help reinforce one's understanding of key concepts.

Now, it is your turn! As you finish this book, you will be equipped and ready to upgrade your license. The path to becoming a General Class License holder is within your reach. This book is your guide, mentor, and companion on this exciting journey.

The airwaves are waiting for you to expand your reach. Welcome to the General Class!

Jared Johnson (KF0RTU/AG) @ Morse Code Publishing

The General Class License Exam

THE GENERAL CLASS LICENSE is your gateway to expanded privileges. With this upgrade, you'll gain access to a wider range of frequencies and operating modes, opening the door to global communication and advanced technical skills.

Earning this license demonstrates your commitment to learning and going beyond the basics of radio. The exam covers a variety of topics, ensuring you're prepared for the challenges and opportunities of being a General Class operator. By passing this exam, you'll not only enhance your skills but also become a more versatile and capable member of the ham radio community.

Chapter 1

Design of this Book

THE AIM OF THIS book is to help you achieve three goals:

1. **Move beyond the basics and expand your knowledge of amateur radio.**

2. **Prepare for and pass the FCC exam to obtain your General license.**

3. **Equip you to expand your reach and connect globally over the airwaves!**

This book is divided into four core sections: The Science of Radio, The Physical Radio Station, Operating Your Radio, and Rules & Regulations. We conclude with a final section on further expanding your skills and knowledge.

Each part of this book forms a critical element in our teaching approach. Together, they work to strengthen your understanding and knowledge of ham radio, leading you toward success.

Rules & Regulations	Science of Radio
The Radio Station	Operating the Radio

The Science of Radio

This section covers the technical side of radio communication. We'll break down essential topics like wave propagation, modulation, and frequency, along with the types of signals and how they move through the atmosphere. These concepts are the building blocks for understanding how radio waves work and how you can use them to communicate.

The Physical Radio Station

Here, we explore the physical setup of your station. From antennas and feedlines to transceivers and power supplies, you'll learn about the key components of a working radio station. We'll also touch on how to design your station and troubleshoot problems so you're always ready to connect with others.

Operating Your Radio

This section is all about hands-on experience. You'll learn to make contacts, use repeaters, and operate in CW, SSB, and digital modes. We'll also cover the basics of clear communication, handling interference, and safety. By the end, you'll be confident in running your station efficiently and responsibly.

Rules & Regulations

This area covers the legal side of amateur radio. We'll review FCC regulations, band plans, licensing, and operator responsibilities. Knowing these rules is critical to ensuring safe and lawful use of the radio spectrum while avoiding interference with other users.

Our Teaching Framework

Our approach to teaching integrates all four sections—Rules & Regulations, The Radio Science, The Physical Radio Station, and Operating Your Radio—so that each part builds on the others. For example, understanding wave propagation will help you improve your station setup and operating skills while knowing the rules ensures you follow proper procedures.

This interconnected method will give you a well-rounded and deeper understanding of amateur radio, ensuring you're ready for the General exam. At the end of the book, you'll find a complete index to each exam question. This will enable you to quickly and easily reference key materials to review as needed.

A quick word of encouragement: expanding your skills in amateur radio can seem like a lot at first, but it's all about focusing on the core ideas. You don't have to memorize everything—just focus on the fundamental principles that drive the field. Once you understand these, you'll see that most other details are variations or applications of these basics.

This book is designed to be straightforward and practical, providing just the information you need without getting bogged down in unnecessary detail. By mastering the core principles and seeing how they fit together, you'll find that what once seemed complicated is now clear and manageable.

Keeping Information Accurate

We've worked hard to make sure the material in this book is accurate and up to date. However, the world of amateur radio continues to evolve, and sometimes mistakes can slip through. If you spot any errors or outdated information, please let us know! Your feedback helps us keep this book current and useful for future readers.

You can reach me at **Jared@MorseCodePublishing.com**.

Thank you for helping us keep this resource valuable for everyone!

Chapter 2

Why Upgrade to General?

Upgrading from the Technician to the General license brings several exciting and practical benefits, significantly expanding an amateur radio operator's capabilities.

One of the most substantial upgrades is access to a broader range of high-frequency (HF) bands, which allows operators to communicate over long distances, including global communication, rather than being mainly limited to local and regional contacts on VHF and UHF.

You'll gain access to portions of popular bands like 80, 40, 20, and 15 meters, which are beneficial for long-distance communication (DXing). But that's not all—you'll also have operating privileges on portions of less commonly used bands such as 160, 60, 30, 17, 12, and 10 meters. This broader spectrum opens up a wide range of frequencies and modes, enhancing your ability to connect with stations across the globe, often under various propagation conditions.

With these additional bands, General license holders can explore broader access with modes such as SSB (Single Sideband), CW (Morse Code), and various digital modes like FT8 and PSK31, which are popular for DX communication.

One of the most compelling reasons to upgrade is the increased power limit. As a General Class licensee, you can legally transmit using up to 1500 watts of PEP (Peak Envelope Power) on most HF bands. This allows you to boost your signal strength, which can be invaluable during challenging propagation conditions or when trying to reach distant stations.

This upgrade also enhances the opportunity to experiment with and fine-tune operating techniques and radio equipment, making it ideal for operators who enjoy the technical aspects of radio. Additionally, access to expanded HF privileges opens doors to participating in more contests, emergency communications, and social networks with amateur operators worldwide, fostering a richer and more versatile ham radio experience.

By upgrading, you'll expand your capabilities in terms of spectrum and power and the variety of ways you can communicate. The General Class license is a gateway to a broader and more exciting amateur radio experience, whether chasing DX, experimenting with digital modes, or enjoying more flexible operating privileges.

Overall, the General license offers both practical tools and a greater sense of freedom on the air.

Onward and upward to General!

Chapter 3

The Exam

IMAGINE THE SENSE OF accomplishment as you key your first transmission using your new General call sign. To reach this milestone, a well-structured and carefully laid out study plan is needed.

Upgrading from a Technician to a General license involves steps designed to ensure you are well-prepared for the additional responsibilities and privileges. The first step is to commit to studying for the General exam, and as you are already reading this part of the book, I think we are good here!

This exam is more challenging than the Technician exam, requiring a deeper understanding of various topics. Consistent study sessions are important and should be scheduled. Allocate specific times each week dedicated to studying different sections of the exam.

Preparing for the Exam

In the first book in this series, *Ham Radio Technician Class License Study Guide: From Beginner to Licensed! Master the Fundamentals of Amateur Radio, Ace the FCC Exam, and Get on the Air with Confidence,* we provided an entire chapter on how to study and feel confident in your study sessions. We will not repeat that here but will provide you with a free downloadable copy of that chapter to use if needed.

How to Study

As you work through this book, treat the questions and answers as part of the regular content—they're integrated in a way to help reinforce your learning. Each question appears alongside four possible answers, with the correct one **bolded** for easy identification. Additionally, important details in the surrounding text will be *underlined*, signaling key information directly related to the exam question.

For instance, take a look at this exam question.

What is a geomagnetic storm? (G3A06)

A. A sudden drop in the solar flux index
B. A thunderstorm that affects radio propagation
C. Ripples in the geomagnetic force
D. A temporary disturbance in Earth's geomagnetic field

Read the question and answers as part of the regular text. They are designed to be read as you find them. Now, focus on the bolded answer and examine the incorrect distractors. Do you understand why they are incorrect? This teaching method will emphasize both the correct answers and the insights provided by the distractors. It's not just about memorizing this one question but also about grasping the core knowledge surrounding it.

Practice Exams

Taking practice exams as part of your study plan is much like rehearsing for a big performance—it offers a critical snapshot of how well you've retained the material and highlights areas that may need more focus. Regularly taking practice exams helps gauge your understanding and prepares you for the exam's structure and timing.

We've included 16 practice exams — yes, 16! Why in the world are there so many? It's important to experience every possible exam question at least once in the format you'll encounter on test day. Section G8C has sixteen potential questions, and since only one will appear on the actual exam, we constructed 16 practice tests to ensure you're ready for whichever one comes your way.

And so be confident. If you can pass these 16 practice exams, the actual exam will be a breeze.

Practice Exams

Scheduling Your Practice Exams

As your exam day draws nearer, try to increase the frequency of your practice exams. In the final weeks before the test, aim to recreate exam conditions as closely as possible. Find a quiet place where you won't be disturbed and time yourself just as you would in the exam setting. This method not only aids in solidifying your knowledge but also helps reduce any nerves by familiarizing you with the pacing and structure of the test. It's also a great way to practice time management, ensuring you can confidently complete all sections within the given timeframe.

The Exam Format

The exam is designed to assess your knowledge across various topics. While it includes similar subject matter to the Technician exam, the General Class questions dive deeper into the technical details, ensuring you're well-prepared for the expanded privileges.

The General Class exam includes 35 multiple-choice questions, and to pass, you'll need to correctly answer at least 26, which is a score of 74%.

The Sub-Elements of the Exam

Understanding how the exam content is weighted helps prioritize your study efforts. The exam sub-elements are weighted differently, which means that some sub-elements will have more questions on the exam than others. This insight directs your focus to areas needing a more robust review.

For instance, you will see two times as many questions about FCC rules (5 questions) as you will for Safety (2). While having well-rounded knowledge is important, emphasizing these heavier-weighted areas can increase your chances of a successful outcome. See below for each sub-element and the number of questions for each sub-element on the exam.

Sub-elements on the General exam:

1. Commission's Rules: (5 questions from this sub-element will be on the exam)

2. Operating Procedures: (5 questions from this sub-element will be on the exam)

3. Radio Wave Propagation (3 questions)

4. Amateur Radio Practices (5 questions)

5. Electrical Principles (3)

6. Circuit Components (2)

7. Practical Circuits (3)

8. Signals and Emissions (3)

9. Antennas and Feed Lines (4)

10. Safety (2)

For a total of 35 exam questions.

Taking Your Exam

We are going to make an assumption at this point. The assumption is that you took and passed your Technician's exam. Therefore, you know about FRN (FCC Registration Number) and how to navigate the FCC Website.

You also should know how to schedule your exam since you've already done it for the Technician's. All this to say, we will list some high-level resources below but won't go into as much detail as we did in the *Technician's Study Guide*. If you do have questions, please reach out, and we would be happy to fill in any details for you, just in case it's been a while since your Technician's exam: **Jared@MorseCodePublishing.com**

Schedule Your Exam

Visit the ARRL website and navigate to the search for exam session page. Register for the exam that works for you.

- https://www.arrl.org/find-an-amateur-radio-license-exam-session

  - Or type into Google, "Find an Amateur Radio License Exam in Your Area," and click on the ARRL site that comes up

Exam Day

Exam day represents the culmination of all your hard work and preparation as you take a significant step toward upgrading your amateur radio license. Focusing on mental and physical readiness is important to perform at your best.

Bring a valid photo ID, such as a driver's license or passport. Minors may need a school ID and possibly a guardian's ID. You'll also need any previous Amateur Radio licenses or Certificates of Successful Completion (CSCE), two pencils, a pen, a calculator with cleared memory, and the exam fee. Cell phones and other electronic devices with calculator functions are prohibited.

This should all be a reminder.

After successfully passing, you'll need to complete and submit the necessary paperwork and fees with assistance from the Volunteer Examiners (VEs). Once everything is processed, your new General Class license will be issued, granting you broader operating privileges.

With that overview out of the way, let's begin studying for your General Class License!

The Science of Radio

THE SCIENCE OF RADIO forms the foundation of everything we do as amateur radio operators. Understanding the principles behind how radio waves travel, interact with the environment, and carry information is essential to becoming a skilled and effective operator. This section dives into the technical concepts that make radio communication possible.

By exploring these core topics, you'll gain a deeper appreciation for the physics and engineering behind the signals you transmit and receive. Whether you're building antennas, optimizing your equipment, or troubleshooting issues, the knowledge in this section will empower you to tackle challenges with confidence and curiosity. Mastering the science of radio is not only a key step in passing the General Class License exam but also in unlocking the full potential of amateur radio.

Chapter 4

Propagation

UNDERSTANDING PROPAGATION IS A cornerstone of effective radio communication. Propagation refers to the way radio waves travel from a transmitter to a receiver, and it's influenced by a variety of factors, including frequency, atmospheric conditions, and the environment.

For General Class license holders, knowing how propagation works is necessary to maximize the reach and reliability of their transmissions. Whether it's bouncing signals off the ionosphere for long-distance communication or using ground waves for local coverage, mastering propagation concepts will empower you to adapt to different operating conditions and make the most of your equipment.

This chapter explores key propagation topics, giving you the knowledge needed to communicate effectively across the bands and pass your exam.

Propagation Types

Line-of-Sight

Line-of-sight propagation is the simplest form, applicable primarily to VHF and UHF communications. These signals travel in a straight line from the transmitter to the receiver, making them ideal for short-range communications. However, obstacles like buildings or hills can block these signals, limiting their range. This mode is commonly used in urban environments and local communications, where clear paths between antennas are maintained.

The Technician's exam largely covered aspects of line-of-sight propagation.

Ground

Ground wave propagation is another particularly relevant mode for HF communications over short distances. These waves follow the contour of the Earth, bending slightly to maintain contact with the ground. Ground waves are most effective at lower frequencies, covering distances up to several hundred miles. This mode is less affected by obstacles than line-of-sight, making it reliable for regional communications. However, ground wave signals gradually lose strength as they travel, and the Earth's curvature and terrain limit their range.

Skywave

Skywave propagation, also known as ionospheric reflection or skip propagation, is the key to long-distance HF communications. In this mode, radio waves are reflected or refracted ("skipped") by the ionosphere, a layer of the Earth's atmosphere filled with charged particles. This natural phenomenon allows signals to travel beyond the horizon, bouncing between the ionosphere and the Earth's surface to cover thousands of miles. Skywave propagation (skip propagation) is highly effective for global communication, enabling contact with stations on different continents.

The distance a signal can "skip" depends on factors like the frequency, the ionosphere layer it bounces off, and the current solar activity. Higher ionospheric layers, like the F2 region, typically enable longer skip distances.

Skip Zone

A transmitting station's "skip zone" is an area surrounding the transmitter where radio signals sent via skywave propagation cannot be received. This happens because the ionosphere refracted the radio waves at distances farther away from the station, skipping over the region closest to the transmitter. The signal is reflected to Earth farther out, leaving a gap between the point where ground wave signals weaken and the area where skywave signals return.

To visualize it, imagine the radio signal bouncing off the ionosphere like a ball. The signal travels upward, skipping over nearby areas and returning to Earth farther away. The skip zone is the area in between—where neither ground waves nor skywave signals reach effectively. Think of this as a kind of no man's land. This can be an issue for stations trying to communicate with nearby locations using higher-frequency signals that rely on ionospheric reflection.

Scatter

However, with all things related to radio waves, it is not quite that simple. Scatter propagation is a type of propagation that allows signals to be heard in the transmitting station's skip zone. Instead of the signal being fully absorbed or reflected to Earth by the ionosphere, a portion of the signal is scattered in different directions by irregularities in the ionosphere or the Earth's atmosphere. This scattering can cause weak signals to reach areas within the skip zone, where normal ground-wave or skywave signals would not be heard.

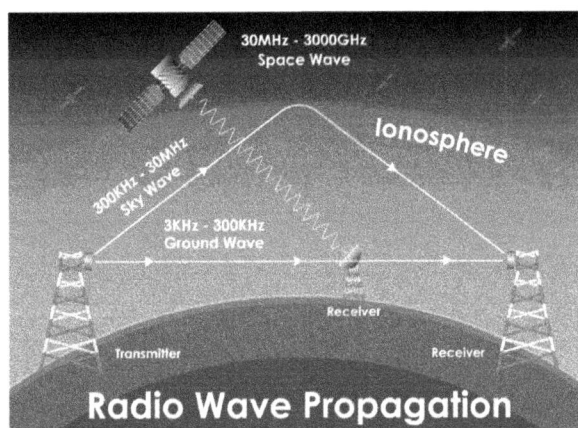

Radio Wave Propagation

Think of scatter propagation as "filling in the gap." It lets signals get through to areas typically skipped over by direct ionospheric reflection, making communication possible even in the zones close to the transmitter that are usually silent.

What type of propagation allows signals to be heard in the transmitting station's skip zone? (G3C09)

A. Faraday rotation
B. Scatter
C. Chordal hop
D. Short-path

HF Scatter

HF scatter signals often have a characteristic fluttering or wavering sound. This occurs because the signal is scattered in multiple directions by irregularities in the ionosphere, causing parts of the signal to take different paths and arrive at slightly different times. The result is a fluctuating, unstable signal that creates a noticeable "flutter" when received.

What is a characteristic of HF scatter? (G3C06)

A. Phone signals have high intelligibility
B. Signals have a fluttering sound
C. There are very large, sudden swings in signal strength
D. Scatter propagation occurs only at night

HF scatter signals sound distorted because the radio wave's energy is scattered into the skip zone through several paths. As the signal encounters irregularities in the ionosphere, parts of it are reflected or refracted in multiple directions. These scattered signal portions take slightly different routes, meaning they arrive at the receiver at different times. This causes the signal to blend together unevenly, resulting in a distorted or fluttering sound.

Scatter propagation creates a "multi-path effect," where the signal is split and reassembled imperfectly at the receiver. This scattering process produces a signal that sounds less clear and distorted than a direct or single-path signal.

What makes HF scatter signals often sound distorted? (G3C07)

A. The ionospheric region involved is unstable
B. Ground waves are absorbing much of the signal
C. The E region is not present
D. Energy is scattered into the skip zone through several different paths

And finally, HF scatter signals in the skip zone are usually weak because only a small portion of the signal's energy is scattered into this area. Instead of the entire signal being reflected to Earth, most of the signal follows the regular skywave propagation path, skipping over the zone. The small amount of energy that does get scattered by irregularities in the ionosphere is spread out in various directions, meaning only a tiny fraction reaches the skip zone, resulting in a much weaker signal.

Why are HF scatter signals in the skip zone usually weak? (G3C08)

A. Only a small part of the signal energy is scattered into the skip zone
B. Signals are scattered from the magnetosphere, which is not a good reflector
C. Propagation is via ground waves, which absorb most of the signal energy
D. Propagation is via ducts in the F region, which absorb most of the energy

Short & Long Path

Short-path and long-path propagation refer to two different routes a radio signal can take when traveling using skywave propagation.

1. **Short-path propagation** is the signal's most direct route between two locations, typically following the shortest distance over the Earth's surface. The signal bounces off the ionosphere and returns to Earth after covering the shortest distance between the transmitter and receiver. For example, suppose you're communicating with someone on the opposite side of the continent. In that case, the signal will likely travel in a short arc across the globe.

2. **Long-path propagation** is when the signal travels the long way around the Earth, taking the opposite direction. Instead of following the shortest route, the signal bounces off the ionosphere. It travels around the other side of the globe before reaching the receiver. This long-path route often takes several times the distance of the short path, which causes the signal to arrive slightly later.

Both types of propagation can happen simultaneously, especially on higher HF bands. When a skywave signal reaches your location through both short-path and long-path propagation, you might hear a slightly delayed echo.

This happens because the signal takes two different routes to get you: the short path, the most direct route, and the long path, which travels around the Earth in the opposite direction. Since the long-path signal has a greater distance to travel, it arrives a fraction of a second later, creating the echo effect.

What is a characteristic of skywave signals arriving at your location by both short-path and long-path propagation? (G3B01)

A. Periodic fading approximately every 10 seconds
B. Signal strength increased by 3 dB
C. The signal might be cancelled, causing severe attenuation
D. A slightly delayed echo might be heard

An easy way to remember this for the exam is to think of the signal as taking "two routes" to get to you. The short path is the quick way, while the long path takes the scenic route around the globe. The result is that slight delay, which you hear as an echo, making this a distinctive characteristic of skywave signals.

Ionospheric Propagation

Picture yourself standing at the edge of a vast ocean. You know there are islands in the distance, but you cannot see them. Instead, you rely on the waves to carry messages across the expanse. In the world of amateur radio, the ionosphere serves a similar purpose. It is an invisible but powerful medium that allows your signals to travel vast distances, bouncing off its layers like stones skipping across water. As you upgrade to a General Class license, understanding ionospheric propagation becomes more essential for optimizing your communication reach, given your expanded HF privileges.

The ionosphere is a region of the Earth's upper atmosphere that overlaps with parts of the mesosphere, thermosphere and exosphere. It is crucial in long-distance radio communication. It is composed of several layers filled with ions and free electrons created by the interaction of solar radiation with atmospheric particles. These layers are not rigid but dynamic, fluctuating in response to solar activity and atmospheric conditions. The ionosphere is typically divided into three primary layers: the D layer, the E layer, and the F layer. Each of these layers has unique characteristics that influence radio wave propagation.

When radio waves have frequencies below the Maximum Usable Frequency (MUF) and above the Lowest Usable Frequency (LUF), the ionosphere refracts or bends them back to Earth. More on MUF and LUF coming up. This allows long-distance communication by "skipping" the signal between the Earth and the ionosphere. The ionosphere acts like a mirror for these frequencies, reflecting them rather than letting them escape into space.

How does the ionosphere affect radio waves with frequencies below the MUF and above the LUF? (G3B05)

A. They are refracted back to Earth
B. They pass through the ionosphere
C. They are amplified by interaction with the ionosphere
D. They are refracted and trapped in the ionosphere to circle Earth

D Layer

The D layer, situated between 50 and 90 kilometers (30-90 miles) above the Earth, is the lowest of the ionospheric layers and closest to the Earth. It forms during the daytime when high-energy X-rays and ultraviolet (UV) light from the Sun ionize atmospheric particles. The D layer is known for its absorbent properties, particularly affecting low-frequency signals. As radio waves pass through this layer, they can experience significant attenuation, reducing their strength.

Which ionospheric region is closest to the surface of Earth? (G3C01)

A. The D region
B. The E region
C. The F1 region
D. The F2 region

During the day, it plays an key role in absorbing low-frequency radio waves, which can weaken or block those signals from traveling far. The D region of the ionosphere is the most absorbent of signals below 10 MHz during daylight hours.

Which ionospheric region is the most absorbent of signals below 10 MHz during daylight hours? (G3C11)

A. The F2 region
B. The F1 region
C. The E region
D. The D region

As a result, long-distance communication during the day on the 40-, 60-, 80-, and 160-meter bands is more difficult because the D region of the ionosphere absorbs signals at these frequencies. The Sun's radiation increases the ionization in the D region, which causes it to absorb lower-frequency/longer-wavelength signals, making it harder for them to travel long distances.

Why is long-distance communication on the 40-, 60-, 80-, and 160-meter bands more difficult during the day? (G3C05)

A. The F region absorbs signals at these frequencies during daylight hours
B. The F region is unstable during daylight hours
C. The D region absorbs signals at these frequencies during daylight hours
D. The E region is unstable during daylight hours

However, the D region almost disappears at night, allowing these signals to travel greater distances.

An easy way to remember this is to think of "D" for "Daytime Absorption." The D region is most active and absorbent when the Sun is out. At night, the D region loses much of its ionization, allowing those same signals to travel farther with less interference. So, if you're trying to communicate on these lower frequencies, nighttime will be your friend!

E Layer

Above the D layer lies the E layer, which extends from about 90 to 120 kilometers (60-70 miles) above the Earth. This layer is less dense than the D layer but still significantly influences radio wave propagation.

When a radio signal bounces off the E region of the ionosphere, it typically covers a distance of up to 1,200 miles (about 1,930 kilometers) in one "hop" before it returns to Earth. The E region can reflect signals in the 3 to 30 MHz range (HF Range). This makes it useful for medium-distance communication, allowing operators to reach stations within a 1,200-mile radius without relying on ground-wave propagation.

> **What is the approximate maximum distance along the Earth's surface normally covered in one hop using the E region? (G3B10)**
>
> A. 180 miles
> **B. 1,200 miles**
> C. 2,500 miles
> D. 12,000 miles

One notable phenomenon associated with the E layer is Sporadic E propagation. Sporadic E occurs when dense patches of ionization form within the E layer, reflecting VHF signals over long distances. This can enable unexpected and fascinating communication opportunities, particularly during the summer months when Sporadic E activity peaks. However, the E layer weakens at night, reducing its impact on signal propagation.

If you aim to cover even greater distances, multiple hops or higher layers like the F region come into play. But for one hop, the E region is perfect for those medium-range contacts!

F Layer

The F layer, located between 240 and 500 kilometers above the Earth (150-300 miles), is the highest and most critical layer for long-distance HF communication. This greater altitude (highest) allows radio signals to travel much farther before bouncing back to Earth, resulting in longer skip distances.

> **Why is skip propagation via the F2 region longer than that via the other ionospheric regions? (G3C03)**
>
> A. Because it is the densest
> B. Because of the Doppler effect
> **C. Because it is the highest**
> D. Because of temperature inversions

During the daytime, the F layer splits into two sub-layers: F1 and F2.

The F1 layer resides at lower altitudes and has moderate ionization levels, while the F2 layer, higher up, has the highest concentration of ions and free electrons.

The F2 region is the highest ionospheric layer. It remains ionized at night, making it ideal for long-distance communication, especially on the higher HF bands (3 to 30 MHz). Signals can reflect off the F2 layer, allowing them to travel thousands of kilometers and enabling global communication. In fact, <u>in one hop, signals can typically cover a distance of up to 2,500 miles</u> (about 4,000 kilometers) along the Earth's surface. Basically, double the distance of the E layer.

What is the approximate maximum distance along the Earth's surface normally covered in one hop using the F2 region? (G3B09)

A. 180 miles
B. 1,200 miles
C. 2,500 miles
D. 12,000 miles

This is why ham radio operators can reach contacts halfway around the world with the help of the F2 layer, particularly during periods of good solar activity. To remember this, think of the "F2" layer as the "Far-reaching" layer, enabling signals to travel great distances, up to 2,500 miles in just one bounce.

Critical Frequency

The term "critical frequency" refers to the <u>highest frequency that can be refracted or bent back to Earth</u> by the ionosphere at a given incidence angle (the angle at which the signal is transmitted). Suppose a signal is transmitted at a frequency higher than the critical frequency. In that case, it will pass through the ionosphere and escape into space instead of being reflected back to Earth.

What is meant by the term "critical frequency" at a given incidence angle? (G3C02)

A. The highest frequency which is refracted back to Earth
B. The lowest frequency which is refracted back to Earth
C. The frequency at which the signal-to-noise ratio approaches unity
D. The frequency at which the signal-to-noise ratio is 6 dB

Think of the "critical frequency" as the upper limit for successful skywave communication at a specific angle. When your transmission frequency is below the critical frequency, the ionosphere will reflect the signal back to Earth. If you go above it, the signal won't return, making it necessary to know this limit for effective long-distance communication.

Critical Angle

The "critical angle" in radio wave propagation refers to <u>the highest angle at which a radio signal can be transmitted and still return to Earth by the ionosphere</u>. Suppose the signal is sent at a higher angle than the critical angle. In that case, it will pass through the ionosphere and not be reflected, effectively disappearing into space. The critical angle is determined by specific ionospheric conditions, including the density and height of the ionized layers, which can vary based on factors like solar activity and time of day.

What does the term "critical angle" mean, as applied to radio wave propagation? (G3C04)

A. The long path azimuth of a distant station
B. The short path azimuth of a distant station
C. The lowest takeoff angle that will return a radio wave to Earth under specific ionospheric conditions
D. The highest takeoff angle that will return a radio wave to Earth under specific ionospheric conditions

---

To determine the critical angle for radio wave propagation, you need to consider two main factors: the frequency of the signal you're transmitting and the ionospheric conditions.

Here's a simplified explanation of how it works:

1. **Ionospheric Density**: The ionosphere's ability to reflect a signal depends on its level of ionization, which varies by time of day, solar activity, and atmospheric conditions. The more ionized the layer, the higher the frequency of signals it can reflect.

2. **Frequency**: The higher the frequency of your signal, the smaller the critical angle. If the signal is transmitted at an angle that is too high, it will escape into space instead of being reflected back to Earth.

Calculating the critical angle requires complex formulas, but in practice, it is often determined through real-time propagation prediction tools or trial and error by adjusting your antenna's takeoff angle and experimenting with different frequencies.

An easy way to remember this is to consider the critical angle as the "ceiling" for successful signal reflection. If you go above that ceiling, your signal escapes the ionosphere. Still, if you stay below it, your signal will bounce back to Earth, allowing you to communicate over long distances.

Maximum & Lowest Usable Frequency

In radio communication, MUF (Maximum Usable Frequency) and LUF (Lowest Usable Frequency) are terms used to describe the limits within which a radio frequency can be successfully transmitted over long distances through the ionosphere.

- **MUF (Maximum Usable Frequency):** This is the highest/ maximum frequency at which a radio wave can be transmitted (used) between two points by refracting off the ionosphere under specific conditions. Signals above the MUF will typically pass through the ionosphere into space rather than being refracted back to Earth. The MUF changes with factors like time of day, season, and solar activity, generally increasing with stronger solar activity.

- **LUF (Lowest Usable Frequency):** This is the lowest frequency that can propagate between two points without significant signal degradation due to ionospheric absorption or other losses. Frequencies below the LUF will often be absorbed in the D-layer of the ionosphere, especially during the daytime. The LUF is affected by atmospheric conditions and can vary based on the same factors as the MUF. However, it tends to be more susceptible to increased ionization and lower at night.

What does MUF stand for? (G3B08)

A. The Minimum Usable Frequency for communications between two points
B. The Maximum Usable Frequency for communications between two points
C. The Minimum Usable Frequency during a 24-hour period
D. The Maximum Usable Frequency during a 24-hour period

What does LUF stand for? (G3B07)

A. The Lowest Usable Frequency for communications between two specific points
B. Lowest Usable Frequency for communications to any point outside a 100-mile radius
C. The Lowest Usable Frequency during a 24-hour period
D. Lowest Usable Frequency during the past 60 minutes

LUF marks the minimum frequency needed to avoid excessive absorption by the ionosphere, ensuring your signal can travel the distance. Staying above the LUF helps to maintain a strong, stable connection.

Radio waves with frequencies below the Lowest Usable Frequency (LUF) are usually attenuated, or weakened before they can reach their intended destination. This happens because lower frequencies are more susceptible to absorption by the ionosphere, particularly in the D layer during the daytime. As a result, these signals lose strength and can fade out before they have the chance to refract back to Earth and complete the long-distance communication path.

What usually happens to radio waves with frequencies below the LUF? (G3B06)

A. They are refracted back to Earth
B. They pass through the ionosphere
C. They are attenuated before reaching the destination
D. They are refracted and trapped in the ionosphere to circle Earth

Frequencies below this threshold get absorbed rather than reflected, which means they won't be able to travel far enough to connect with distant stations. Staying above the LUF ensures that your signal is strong enough to overcome ionospheric absorption and reach its destination.

MUF (Maximum Usable Frequency) and LUF (Lowest Usable Frequency) guide operators in selecting the ideal frequency range for long-distance HF communication. Staying within this "sweet spot" ensures signals are refracted back to Earth rather than escaping into space (above the MUF) or being absorbed by the atmosphere (below the LUF). This balance is key to reliable, effective communication.

Factors Impacting MUF

The MUF is influenced by several key factors: path distance and location, time of day and season, and solar radiation and ionospheric disturbances.

- **Path distance and location** affect the MUF because longer paths generally require higher frequencies to maintain a connection, as the ionosphere needs to refract the signal over a greater distance.

- **Time of day and season** are also important—during daylight hours and summer months, increased solar radiation causes more ionization in the ionosphere, which typically raises the MUF. Conversely, ionization decreases at night and in winter, often lowering the MUF.

- **Solar radiation and ionospheric disturbances** play a significant role, too. High solar activity, such as sunspots and solar flares, increases ionization and can raise the MUF, while disturbances like geomagnetic storms can disrupt the ionosphere and reduce the MUF.

What factors affect the MUF? (G3B02)

A. Path distance and location
B. Time of day and season
C. Solar radiation and ionospheric disturbances
D. All these choices are correct

Remember MUF as being shaped by the "when" (time and season), the "where" (path and location), and the "what" (solar activity) of your signal path. Together, these elements determine the highest frequency that will reliably refract back to Earth for effective communication.

How to Determine MUF & LUF

For long-distance skip propagation, a frequency just below the Maximum Usable Frequency (MUF) will experience the least attenuation. The MUF represents the highest frequency that can be reliably refracted back to Earth by the ionosphere. Using a frequency just below the MUF ensures that the signal will be strong enough to avoid excessive absorption and reach distant stations with minimal loss of strength.

Which frequency will have the least attenuation for long-distance skip propagation? (G3B03)

A. Just below the MUF
B. Just above the LUF
C. Just below the critical frequency
D. Just above the critical frequency

By operating just below this boundary, you're optimizing your signal for minimal attenuation, ensuring it can travel long distances with reduced fading and signal loss. This choice of frequency helps maintain a clear, stable connection over vast areas.

To figure out the Maximum Usable Frequency (MUF) and Lowest Usable Frequency (LUF), operators rely on a combination of propagation tools, real-time data, and practical experience. Here's a simplified process:

Determining MUF

1. **Ionospheric Data**: Operators use online resources like ionospheric maps, space weather updates, or tools such as the VOACAP (Voice of America Coverage Analysis Program) to check real-time ionospheric conditions.

2. **Calculation or Software**: MUF is typically calculated based on the critical frequency (the highest frequency that will reflect at vertical incidence) and the transmission angle. Many propagation prediction tools automatically perform this calculation.

3. **Trial and Error**: Operators can test various frequencies within their band privileges to find the highest frequency that supports reliable communication over the intended path.

Determining LUF

1. **Atmospheric Noise and Absorption**: The LUF is influenced by atmospheric conditions, signal strength, and noise levels. Operators look at propagation tools and band reports to assess how lower frequencies are performing.

2. **Trial and Error**: Testing the lower frequencies and monitoring signal quality can help determine the point below which communication becomes unreliable.

Practical Tools

- **Software Solutions**: Tools like VOACAP, HamCAP, or even smartphone apps provide predictions for MUF and LUF based on input parameters like time, date, location, and solar activity.

- **Online Resources**: Websites and live propagation maps offer up-to-date MUF and LUF values, as well as solar indices like the K-index or solar flux, which affect propagation.

By combining these tools with experience, operators can effectively select the optimal frequencies for communication under varying propagation conditions.

LUF exceeds MUF

<u>When the Lowest Usable Frequency (LUF) exceeds the Maximum Usable Frequency (MUF), ordinary skywave communication over that path becomes impossible.</u> This situation occurs because the ionosphere can no longer support the reflection of any frequencies within the usable range. The ionosphere absorbs signals below the LUF, and signals above the MUF pass through it without being refracted back to Earth, leaving no viable frequencies for long-distance communication.

What happens to HF propagation when the LUF exceeds the MUF? (G3B11)

A. Propagation via ordinary skywave communications is not possible over that path
B. HF communications over the path are enhanced
C. Double-hop propagation along the path is more common
D. Propagation over the path on all HF frequencies is enhanced

---

To remember this for the exam, think of the LUF and MUF as the boundaries of a usable "window" for HF propagation. If the LUF rises above the MUF, this "window" closes, preventing effective skywave communication on that path. In such cases, operators might need to wait until conditions change and the LUF drops below the MUF to restore reliable HF propagation.

When could this happen?

The situation where the Lowest Usable Frequency (LUF) exceeds the Maximum Usable Frequency (MUF) typically occurs during severe ionospheric disturbances or under extreme conditions that disrupt normal skywave propagation. Here are some specific scenarios when this might happen:

1. Severe Solar Storms

- During a major solar flare or coronal mass ejection (CME), the D-layer of the ionosphere becomes highly ionized, drastically increasing absorption of lower frequencies. This causes the LUF to rise significantly.

- Simultaneously, the upper ionospheric layers (E and F layers) may become unstable or depleted, reducing the MUF as they lose their ability to refract higher frequencies.

2. Geomagnetic Storms

- Intense geomagnetic activity can disrupt the ionosphere's structure, creating turbulent and irregular ionization patterns. This can cause both higher absorption (raising the LUF) and a loss of higher-frequency reflectivity (lowering the MUF).

3. Nighttime on Low-Solar-Activity Cycles

- During periods of very low solar activity (solar minimum), combined with nighttime conditions, the ionosphere becomes weakly ionized. The MUF drops significantly as the F-layer loses its ability to reflect higher frequencies, while increased D-layer absorption raises the LUF.

4. Polar Regions During Auroral Events

- In polar regions, auroral events caused by charged particles from the sun can severely disrupt the ionosphere. This leads to extreme absorption at lower frequencies (high LUF) while destabilizing higher-frequency propagation (low MUF).

Practical Example:

Imagine you're trying to use HF communication during a solar storm. The usual bands, like 20 meters, which might normally support propagation for long-distance communication, fail because the LUF has risen to, say, 18 MHz, while the MUF has dropped to 12 MHz. There is no frequency band in the HF range that can propagate via skywave, effectively blocking communication on that path.

Understanding these conditions allows operators to switch to alternative methods, such as VHF/UHF communication or digital modes that rely on internet-linked systems during such challenging times.

Near Vertical Incidence Skywave

Near Vertical Incidence Skywave (NVIS) propagation is a method of using medium-frequency (MF) or high-frequency (HF) radio waves to communicate over short distances by transmitting signals at high elevation angles—almost straight up into the sky. These signals bounce off the ionosphere and are reflected back to Earth, covering areas within a range of about 200 to 400 miles (300 to 600 kilometers). This technique is instrumental in hilly or mountainous regions where direct line-of-sight communication isn't possible.

NVIS propagation is closely related to the concept of critical angle. In NVIS, radio signals are transmitted to the ionosphere at a very high angle—almost straight up. For the signal to be refracted back to Earth (rather than passing through the ionosphere into space), the frequency must be below the critical frequency, and the angle of transmission must be lower than the critical angle.

What is near vertical incidence skywave (NVIS) propagation? (G3C10)

A. Propagation near the MUF
B. Short distance MF or HF propagation at high elevation angles
C. Long path HF propagation at sunrise and sunset
D. Double hop propagation near the LUF

The critical angle determines the maximum angle at which a signal can still be refracted by the ionosphere and returned to Earth. For NVIS to work, the signal is transmitted at a near-vertical angle below the critical angle. This ensures the signal is bent back down over a short distance, enabling effective local communication. So, NVIS relies on using frequencies low enough (below the critical frequency) to allow this high-angle reflection.

Solar Activity

The Sun and its activities play an interesting role in radio communication, particularly for operators using HF bands. The Sun constantly emits energy in the form of electromagnetic radiation and charged particles, which interact with Earth's atmosphere and magnetic field. This solar energy, especially during periods of high activity like solar flares and increased sunspot numbers, enhances the ionosphere's ability to refract radio waves to Earth, making long-distance communication possible.

However, solar activity can be a double-edged sword. While it can improve signal propagation, intense solar events like geomagnetic storms can also disrupt communications.

Solar Cycle

The solar cycle spans approximately 11 years and alternates between solar maximum and minimum periods. During solar maximum, increased solar activity boosts ionization, while during solar minimum, activity is reduced and propagation conditions are weaker.

The cycle begins at a solar minimum when the Sun has few or no sunspots and low solar activity. It gradually builds up to a solar maximum, where sunspot numbers and solar activity peak. After the maximum, the activity declines again, leading to the following solar minimum.

The solar cycle is crucial for radio operators because solar activity directly affects ionospheric conditions, which influences HF radio wave propagation. During high solar activity (solar maximum) periods, more sunspots increase ionization in the ionosphere, improving HF propagation, especially on higher frequencies. Conversely, lower solar activity can lead to poorer HF propagation during solar minimums, particularly at higher frequencies.

As of January 2025, we are in Solar Cycle 25, which began in December 2019. Solar cycles typically last about 11 years, with activity peaking around the midpoint. Predictions indicate that Solar Cycle 25 will reach its maximum in July 2025.

Increased solar activity during this period can enhance ionospheric propagation, benefiting long-distance radio communications. However, it also raises the likelihood of solar flares and geomagnetic storms, which can disrupt communications and affect satellite operations. Staying informed about solar activity is crucial for radio operators to optimize performance and mitigate potential disruptions.

Exam question time.

The 20-meter band (14 MHz) typically supports worldwide propagation during daylight hours at any point in the solar cycle. The 20-meter band is well-suited for long-distance communication, regardless of whether solar activity is high or low. While higher solar activity (during solar maximum) enhances propagation on many bands, the 20-meter band remains reliable for worldwide communication even during solar minimum, making it a popular choice for amateur operators seeking consistent long-distance contacts.

At what point in the solar cycle does the 20-meter band usually support worldwide propagation during daylight hours? (G3A07)

A. At the summer solstice
B. Only at the maximum point
C. Only at the minimum point
D. At any point

Remember, the solar cycle is a natural rhythm of the Sun's activity that peaks and wanes every 11 years, and this cycle heavily impacts radio communication conditions.

28 Day Cycle

HF propagation conditions vary in a 26- to 28-day cycle due to the Sun's rotation around its axis. The Sun rotates once approximately every 27 days, causing different features on its surface, such as sunspots and coronal holes, to periodically face Earth. As these solar features come into view, they affect Earth's ionosphere by altering how radio waves are refracted, resulting in changes to HF propagation. For instance, when an active region with sunspots or a coronal hole rotates into view, it can either enhance or disrupt HF communication, depending on the solar activity.

What causes HF propagation conditions to vary periodically in a 26- to 28-day cycle? (G3A10)

A. Long term oscillations in the upper atmosphere
B. Cyclic variation in Earth's radiation belts
C. Rotation of the Sun's surface layers around its axis
D. The position of the Moon in its orbit

Think of the Sun as a rotating "influencer" that regularly changes the conditions we experience on Earth. As the Sun's surface turns, different areas emit solar wind and radiation that impact HF propagation, leading to a repeating cycle of conditions roughly every 27 days.

Understanding the impact of solar activity helps radio operators predict and adapt to changing conditions for optimal performance. Also, if you follow any Ham Radio groups on Facebook, there is always lots of chatter and excitement around the Sun's activities.

Solar Flares

Solar flares are a sudden burst of intense radiation. These flares can cause sudden ionospheric disturbances (SIDs), leading to short-term radio fade-outs or enhanced absorption of electromagnetic waves.

The increased ultraviolet (UV) and X-ray radiation from a solar flare affects radio propagation on Earth in approximately 8 minutes. This is because it takes about 8 minutes for the radiation from the Sun to travel 93 million miles to Earth, moving at the speed of light. Once this radiation reaches our planet, it can immediately cause disturbances in the ionosphere, particularly on the sunlit side of Earth, leading to short-term radio signal blackouts or fading.

Approximately how long does it take the increased ultraviolet and X-ray radiation from a solar flare to affect radio propagation on Earth? (G3A03)

A. 28 days
B. 1 to 2 hours
C. 8 minutes
D. 20 to 40 hours

Solar flares send bursts of intense radiation that disrupt radio communications quickly after they happen, giving operators little time to prepare once the flare occurs.

Impact from Sudden Ionospheric Disturbances

A sudden ionospheric disturbance (SID) disrupts daytime ionospheric propagation, especially affecting lower-frequency signals. When a SID occurs—often triggered by a solar flare—it floods the ionosphere with intense ultraviolet and X-ray radiation. This extra energy causes increased ionization, particularly in the D layer, which absorbs lower-frequency HF signals and weakens or blocks them from being refracted back to Earth.

What effect does a sudden ionospheric disturbance have on the daytime ionospheric propagation? (G3A02)

A. It enhances propagation on all HF frequencies
B. It disrupts signals on lower frequencies more than those on higher frequencies
C. It disrupts communications via satellite more than direct communications
D. None, because only areas on the night side of the Earth are affected

While higher frequencies may still get through, lower frequencies are much more likely to be absorbed, disrupting long-distance communication on those bands during the day.

Sunspots

A sunspot is a temporary dark spot that appears on the surface of the Sun due to lower temperatures compared to the surrounding areas. These spots are caused by intense magnetic activity, which inhibits the transfer of heat to the surface, making them cooler and darker than the surrounding solar material. While they may look small in images, sunspots can be massive, often much larger than Earth.

Sunspots are important to radio operators because they are closely linked to solar activity, which can enhance ionospheric conditions for radio wave propagation. A higher sunspot number generally improves HF propagation, particularly at higher frequencies. Sunspots are indicators of increased solar activity, which boosts the amount of solar radiation reaching Earth's ionosphere. This extra energy causes more ionization in the ionospheric layers, allowing higher-frequency radio waves (such as those on the 10, 12, and 15-meter bands) to be reflected to Earth more efficiently. As a result, long-distance communication on these higher frequencies becomes more reliable and effective.

How does a higher sunspot number affect HF propagation? (G3A01)

A. Higher sunspot numbers generally indicate a greater probability of good propagation at higher frequencies
B. Lower sunspot numbers generally indicate greater probability of sporadic E propagation
C. A zero sunspot number indicates that radio propagation is not possible on any band
D. A zero sunspot number indicates undisturbed conditions

Conversely, low sunspot numbers result in reduced ionization, negatively affecting long-distance communication. So, while sunspots may look dark and calm, they are actually dynamic regions driving much of the sun's activity that directly impacts amateur radio propagation.

Coronal Mass Ejection

A Coronal Mass Ejection (CME) is a large burst of plasma and magnetic field from the Sun's corona, the outermost layer of the Sun's atmosphere. During a CME, billions of tons of charged particles are ejected into space, traveling at high speeds. When these charged particles interact with Earth's magnetic field, they can cause geomagnetic storms, which can impact radio communications, satellite operations, and power grids.

For radio operators, CMEs are significant because they can disrupt HF radio signals by disturbing the ionosphere, leading to degraded signal quality or complete signal loss. Conversely, CMEs can also produce auroras, enhancing VHF signal propagation for short periods by providing an additional reflective surface for signals.

A CME takes much longer than a solar flare to affect radio propagation on Earth, usually between 15 hours and several days. CMEs are massive clouds of plasma and magnetic fields ejected from the Sun, and they travel much slower than the radiation from a solar flare. When a CME reaches Earth, it can cause significant disturbances in the geomagnetic field.

How long does it take a coronal mass ejection to affect radio propagation on Earth? (G3A11)

A. 28 days
B. 14 days
C. 4 to 8 minutes
D. 15 hours to several days

A CME is a "slow but powerful storm" compared to the almost immediate impact of a solar flare. While solar flare effects are felt in minutes, CMEs take much longer to travel from the Sun to Earth, but their impacts can last longer and be more widespread once they arrive.

Solar Flare vs Coronal Mass Ejection

A coronal mass ejection (CME) is different from a solar flare. However, they both originate from the Sun and can affect radio communications on Earth.

- **Solar Flare:** A solar flare is a sudden, intense burst of energy and radiation from the Sun's surface, mainly in the form of electromagnetic waves like X-rays and ultraviolet light. Their impact is felt within 8 minutes. Solar flares can affect radio propagation almost immediately, especially on Earth's daylit side, causing disturbances in the ionosphere. This can lead to short-term radio signal blackouts on HF bands, known as radio blackouts or fade-outs.

- **Coronal Mass Ejection (CME)**: Conversely, a CME involves the ejection of large amounts of plasma and magnetic fields into space. Unlike solar flares, CMEs take longer (15 hours to several days) to reach Earth, but they can cause geomagnetic storms when they do. These storms affect radio communications and satellite operations and can also cause auroras, which sometimes improve VHF propagation.

In summary, solar flares impact radio signals quickly and briefly. CMEs have a delayed but often longer-lasting effect due to their impact on Earth's magnetic field. Both disrupt HF communications but in different ways.

Coronal Holes

A solar coronal hole is a region on the Sun's corona (the outer atmosphere) where the Sun's magnetic field is open, allowing solar wind to escape more readily into space. When viewed in ultraviolet or X-ray light, these regions appear darker than the surrounding areas because they are cooler and less dense. Coronal holes are sources of high-speed solar wind streams, which can travel faster than the solar wind from other parts of the Sun.

When charged particles from solar coronal holes reach Earth, they can disturb HF communication. The high-speed solar wind released from these coronal holes interacts with Earth's magnetic field, causing geomagnetic disturbances. These disturbances disrupt the ionosphere. As a result, long-distance HF communication can become unreliable, with signals fading or wholly blocked.

How is long distance radio communication usually affected by the charged particles that reach Earth from solar coronal holes? (G3A14)

A. HF communication is improved
B. HF communication is disturbed
C. VHF/UHF ducting is improved
D. VHF/UHF ducting is disturbed

The charged particles they send toward Earth create geomagnetic disturbances that make it harder for radio signals to travel long distances, leading to poor HF propagation until the disturbance passes. Unlike solar flares or coronal mass ejections, the effects of coronal holes are usually less intense but can last longer.

Geomagnetic Storms

What is a geomagnetic storm? (G3A06)

A. A sudden drop in the solar flux index
B. A thunderstorm that affects radio propagation
C. Ripples in the geomagnetic force
D. A temporary disturbance in Earth's geomagnetic field

A geomagnetic storm is a temporary disturbance in Earth's geomagnetic field caused by solar activity like solar flares or coronal mass ejections (CMEs). When these bursts of solar energy reach Earth, they interact with our planet's magnetic field, disrupting it.

Just like a thunderstorm can disrupt your day, a geomagnetic storm disrupts radio signals by causing disturbances in the ionosphere, making communication more challenging until the storm subsides.

A geomagnetic storm can significantly degrade HF propagation, especially in high-latitude regions. When solar activity disturbs Earth's geomagnetic field, it disrupts the ionosphere, which is essential for reflecting HF radio signals back to Earth. This disturbance can cause HF signals to fade, become unreliable, or even be blocked entirely, particularly in areas closer to the poles with the strongest geomagnetic effects.

High-latitude operators may experience poor or no signal reception during these storms because the ionosphere becomes too unstable to effectively reflect HF signals to Earth.

How can a geomagnetic storm affect HF propagation? (G3A08)

A. Improve high-latitude HF propagation
B. Degrade ground wave propagation
C. Improve ground wave propagation
D. Degrade high-latitude HF propagation

On the other hand, high geomagnetic activity can benefit radio communications by creating auroras, which can reflect VHF signals. During intense solar activity, such as a geomagnetic storm, charged particles from the Sun interact with Earth's magnetic field and atmosphere, creating beautiful auroras (northern and southern lights). These auroras can act like mirrors in the sky, reflecting VHF signals and allowing communication over longer distances than usual. Hopefully you remember this piece from the tech exam. Or maybe you even tried it!

While geomagnetic storms often disrupt HF communications, they can enhance VHF communications by temporarily allowing signals to reflect off the ionized particles in the auroras, making long-distance VHF contacts possible.

How can high geomagnetic activity benefit radio communications? (G3A09)

A. Creates auroras that can reflect VHF signals
B. Increases signal strength for HF signals passing through the polar regions
C. Improve HF long path propagation
D. Reduce long delayed echoes

Measuring Earth's Geomagnetic Field

The A-index and K-index are two key measurements that help radio operators understand the stability of Earth's geomagnetic field and how it affects radio wave propagation. While both indices measure geomagnetic activity, they do so over different time frames.

The K-index focuses on short-term changes, providing a snapshot of geomagnetic disturbances over a 3-hour period, which helps track immediate conditions that might disrupt HF communications.

It is a measurement of the short-term stability of Earth's geomagnetic field. It ranges from 0 to 9 and indicates how much Earth's magnetic field is being disturbed by solar activity. A low K-index (0-1) means the geomagnetic field is calm and stable. At the same time, a higher K-index (5 or above) indicates significant disturbances, such as those caused by solar storms. High K-index values can lead to disruptions in HF communications and can also affect satellite systems.

What does the K-index measure? (G3A12)

A. The relative position of sunspots on the surface of the Sun
B. The short-term stability of Earth's geomagnetic field
C. The short-term stability of the Sun's magnetic field
D. The solar radio flux at Boulder, Colorado

In contrast, the A-index offers a longer-term view, averaging geomagnetic activity over an entire 24-hour period to show the overall stability of the Earth's magnetic field.

The A-index measures the long-term stability of Earth's geomagnetic field. Unlike the K-index, which tracks short-term fluctuations, the A-index provides an average value over 24 hours to show the overall level of geomagnetic activity. It is expressed on a scale from 0 to 400, with lower numbers indicating stable conditions and higher numbers reflecting more significant disturbances. A low A-index means conditions are calm, which is ideal for radio propagation. At the same time, a high A-index suggests ongoing geomagnetic instability, which can degrade HF communication.

What does the A-index measure? (G3A13)

A. The relative position of sunspots on the surface of the Sun
B. The amount of polarization of the Sun's electric field
C. The long-term stability of Earth's geomagnetic field
D. The solar radio flux at Boulder, Colorado

These indices give radio operators a comprehensive picture of geomagnetic conditions, helping them anticipate short-term and long-term effects on signal propagation.

Low Solar Activity

The 15-meter, 12-meter, and 10-meter bands are the least reliable for long-distance communications during periods of low solar activity. These higher-frequency HF bands depend heavily on strong ionization in the ionosphere to reflect signals to Earth for global communication. During low solar activity, like at the solar minimum, less solar radiation reaches the ionosphere, reducing its ability to reflect signals at these higher frequencies. As a result, signals on these bands are more likely to pass through the ionosphere instead of being refracted back to Earth, making them unreliable for long-distance contact.

Which of the following are the least reliable bands for long-distance communications during periods of low solar activity? (G3A04)

A. 80 meters and 160 meters
B. 60 meters and 40 meters
C. 30 meters and 20 meters
D. 15 meters, 12 meters, and 10 meters

Think of these bands as "high-frequency fair-weather friends." They perform best when solar activity is high, and the ionosphere is energized. Still, they struggle to support long-distance propagation during quiet periods, making lower-frequency bands like 20 meters more dependable in those conditions.

Solar Flux Index (SFI)

The Solar Flux Index (SFI) is a measure of solar radiation at a wavelength of 10.7 centimeters (about 2800 MHz), which is emitted by the Sun's outer layers. This index is used as an indicator of solar activity, as higher values generally correlate with increased ionization in the ionosphere. Higher SFI readings often mean better HF propagation, mainly on higher frequencies, since the increased solar radiation boosts the ionosphere's ability to reflect radio waves.

What is the solar flux index? (G3A05)

A. A measure of the highest frequency that is useful for ionospheric propagation between two points on Earth
B. A count of sunspots that is adjusted for solar emissions
C. Another name for the American sunspot number
D. A measure of solar radiation with a wavelength of 10.7 centimeters

---

Think of the Solar Flux Index as a "solar energy gauge." When the SFI is high, it signals that the Sun is more active, which typically enhances HF communication conditions by energizing the ionosphere. Monitoring the SFI helps radio operators predict when conditions will be favorable for long-distance communication.

Why did the 10.7 cm get called out? The 10.7-centimeter wavelength was specifically chosen for the Solar Flux Index because it is a reliable and consistent indicator of solar activity that correlates well with the levels of ionizing radiation impacting the ionosphere. The Sun's outer layers emit this particular wavelength and is relatively unaffected by the Earth's atmosphere, making it ideal for monitoring.

Scientists found that measuring solar emissions at 10.7 centimeters provided a stable and accurate representation of solar activity, especially sunspot activity. It directly reflects changes in the Sun's magnetic activity, making it a valuable tool for predicting HF radio propagation conditions. The 10.7-centimeter measurement has been used since the 1940s. It is a standard in solar monitoring, helping radio operators and scientists gauge solar conditions that impact communication.

Seasonal Impacts

Seasonal variations also affect propagation conditions. Higher temperatures and increased atmospheric activity can enhance VHF and UHF propagation through phenomena like tropospheric ducting in the summer.

However, during the summer, HF frequencies often experience high atmospheric noise or static levels. This noise is mainly caused by natural events like thunderstorms, which are more frequent in summer, particularly in tropical and subtropical regions. The lightning discharges create static that gets picked up by radio receivers, making it harder to hear signals clearly on these lower frequencies.

Which of the following is typical of the lower HF frequencies during the summer? (G3B12)

A. Poor propagation at any time of day
B. World-wide propagation during daylight hours
C. Heavy distortion on signals due to photon absorption
D. High levels of atmospheric noise or static

The more storms and atmospheric disturbances, the more static you'll likely encounter. This makes communication on lower HF bands noisier during the summer months compared to other times.

With their lower temperatures and reduced atmospheric turbulence, winter months can improve HF propagation, making long-distance contacts more reliable. Understanding these seasonal patterns helps operators plan their activities and maximize communication opportunities.

Determining Propagation Conditions

Propagation conditions frequently change based on atmospheric factors like solar activity and ionospheric conditions. Knowing how signals will behave on a specific band helps operators choose the best frequency for their transmissions. There are several ways to determine current propagation conditions, but one of the most reliable ways <u>to determine current propagation conditions is by using a network of online automated receiving stations.</u>

These networks, such as the Reverse Beacon Network (RBN) or PSKReporter, automatically listen for signals from stations like yours and provide real-time feedback on where your transmissions are being received around the world. By sending a short transmission, you can quickly determine how well your signal propagates and where it's reaching. This method gives you an immediate and reliable indication of propagation conditions on the desired band, making it easier to decide whether to continue using that band or switch to another for better coverage.

Which of the following is a way to determine current propagation on a desired band from your station? (G3B04)

A. Use a network of automated receiving stations on the internet to see where your transmissions are being received
B. Check the A-index
C. Send a series of dots and listen for echoes
D. All these choices are correct

These tools allow you to optimize your radio operations based on current conditions, ensuring your transmissions reach their intended destinations.

Signal Reports

Exchanging signal reports is standard practice at the start of an HF (high frequency) contact in amateur radio. These reports provide a quick assessment of signal quality between stations. This exchange helps operators adjust their equipment, such as transmitter power or antenna orientation, to improve communication-based on current conditions. The reports usually follow the RST system, which evaluates Readability, Signal strength, and Tone, allowing both operators to optimize their setup during the contact.

Why are signal reports typically exchanged at the beginning of an HF contact? (G2D11)

A. To allow each station to operate according to conditions
B. To be sure the contact will count for award programs
C. To follow standard radiogram structure
D. To allow each station to calibrate their frequency display

Signal Reports in HF Contacts

<u>Signal reports are typically exchanged at the beginning of an HF contact to ensure that both stations receive each other clearly and adapt to the prevailing propagation conditions.</u> By sharing a report using the RST (Readability, Signal Strength, and Tone) system, operators can assess how well their signals are being received. This helps them decide whether to adjust transmitter power, reposition their antennas, or modify other technical settings. For example, if the signal strength is weak, an operator might increase power or adjust their antenna to improve reception. This simple exchange helps ensure that the rest of the communication proceeds smoothly.

Chapter Summary

Understanding radio wave propagation is necessary for maximizing your ham radio experience as a General license holder. Propagation affects everything from local contacts to long-distance (DX) communication. It can vary with the time of day, season, solar activity, and the frequency band you use. By learning how factors like the ionosphere, tropospheric effects, and even ground reflections influence your signal, you gain the skills to select optimal frequencies and operating modes.

With this knowledge, you can maximize your reach, anticipate propagation conditions, and make informed choices for effective communication across the bands. Armed with these insights, your ability to adapt and experiment will grow, deepening your understanding and enjoyment of radio.

Chapter 5

Modulation and Bandwidth

MODULATION IS THE PROCESS of encoding information, such as voice or data, onto a carrier wave, allowing it to travel through the atmosphere. Different modulation types have unique characteristics and applications, like AM, FM, and digital methods.

Bandwidth, the range of frequencies a signal occupies, is closely tied to modulation, as each type requires different amounts of bandwidth to transmit clearly and efficiently.

Together, modulation and bandwidth form the backbone of signal quality, efficiency, and reach, making them critical concepts for every radio operator. This chapter explores how these elements work and interact.

Modulation

Modulation is the process of varying a carrier signal to transmit information such as voice, data, or video. At its core, modulation involves altering one or more properties of a carrier wave—typically its amplitude, frequency, or phase—to encode the information you wish to send. The primary purpose of modulation is to make the signal compatible with the transmission medium, allowing it to travel over long distances and through various environments. Without modulation, the raw signal would be too weak or unstable to reach its destination reliably.

You already learned a lot about basic modulation to obtain your Technician's license. For the General test, we will assume you have a baseline foundation of modulation and dive right into the deep end of the modulation pool.

Amplitude Modulation

Amplitude Modulation (AM) is one of the earliest and most straightforward methods of modulating a radio frequency (RF) signal. In AM, the carrier wave's amplitude (or strength) is varied in proportion to the amplitude of the input signal, such as voice or music. The carrier's frequency remains constant, but its instantaneous power level fluctuates based on the information being transmitted. AM has been widely used in broadcasting, particularly for AM radio stations (I'm shocked!) and in some types of voice communication systems. While it is more susceptible to noise and interference than other modulation types, it is still important in radio communication.

Generating an AM signal begins with the modulator circuit, which combines the audio signal with the carrier wave. This modulated signal then passes through an amplifier to boost its strength before being sent to the antenna and transmitted. On the receiving end, the AM receiver uses a demodulator circuit to extract the audio signal from the modulated carrier. This involves rectifying the signal to remove the negative half of the waveform, followed by filtering to remove the high-frequency carrier, leaving only the original audio.

WAVE SHAPES

Analog data signal

Digital data signal

Amplitude modulation (AM)

Digital AM

Frequency modulation (FM)

Carrier signal

What type of modulation varies the instantaneous power level of the RF signal? (G8A05)

A. Power modulation
B. Phase modulation
C. Frequency modulation
D. Amplitude modulation

Remember amplitude modulation as controlling the "power level" or strength of the signal. As the signal's power fluctuates based on the input, the information is embedded in the varying amplitude of the carrier wave. This contrasts with frequency modulation, where the frequency of the carrier is varied instead of its amplitude.

Modulation Envelopes

In AM, a modulation envelope is a key concept to understanding how information is transmitted. The envelope represents the overall shape of the modulated signal, showing how the amplitude of the carrier wave changes in response to the input signal. By visualizing the envelope, you can see the peaks and troughs that correspond to the variations in the signal's amplitude, which carry the information being transmitted. This helps operators monitor signal quality and ensure proper modulation levels.

The modulation envelope of an AM signal is the waveform created by connecting the peak values of the modulated signal. Basically, the signal's outline or "shape" shows how the carrier wave's amplitude varies over time as it carries the information. The carrier itself oscillates rapidly at its own frequency. Still, the envelope gives a clear picture of how the signal's strength changes based on the input being transmitted.

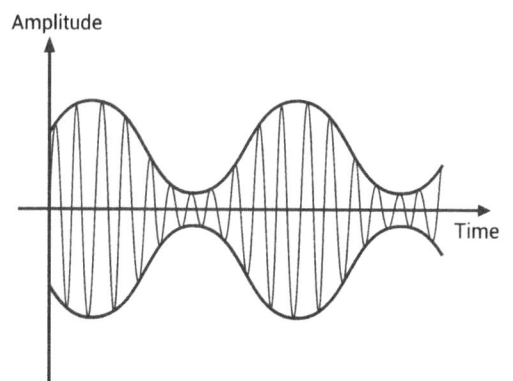

What is the modulation envelope of an AM signal? (G8A11)

A. The waveform created by connecting the peak values of the modulated signal
B. The carrier frequency that contains the signal
C. Spurious signals that envelop nearby frequencies
D. The bandwidth of the modulated signal

The envelope is a visual representation that helps you see how well the signal is being modulated. A clean, undistorted envelope indicates proper modulation, while a distorted or clipped envelope may indicate over-modulation or flat-topping.

Signal Distortion in AM

With AM, the goal is to transmit a clear signal by varying the amplitude of the carrier wave in direct relation to the input. However, when the input levels (such as the drive or speech levels) are too high, it can lead to signal distortion. Distortion not only reduces the quality of the signal but can also cause interference with adjacent frequencies, leading to poor communication.

Flat-topping occurs when an amplitude-modulated phone/ voice signal becomes distorted due to excessive drive or speech levels. When the signal is overdriven, the transmitter cannot accurately follow the peaks of the modulation, causing the signal to "clip" or flatten at the top rather than smoothly following the waveform. This results in a distorted signal, which not only reduces the clarity of the transmitted voice or data but can also cause interference to other stations.

What is meant by the term "flat-topping," when referring to an amplitude-modulated phone signal? (G8A10)

A. Signal distortion caused by insufficient collector current
B. The transmitter's automatic level control (ALC) is properly adjusted
C. Signal distortion caused by excessive drive or speech levels
D. The transmitter's carrier is properly suppressed

Flat-topping "clips" the peaks of the signal due to overdriving the transmitter. The solution is to reduce the drive or speech levels to ensure the transmitter can handle the full range of the signal without distortion.

To prevent overdriving, carefully manage input levels to ensure they stay within the transmitter's linear range. Properly adjusting the audio or modulating input levels helps keep signal integrity and avoid distortion. Using a modulation monitor allows operators to confirm that the transmitter is not exceeding 100% modulation. Additionally, ensuring that microphone and preamp gain settings are correctly adjusted will further reduce the risk of overdriving. By following these practices, operators can maintain clean and compliant transmissions, free from distortion or interference.

Frequency Modulation

Frequency Modulation (FM) works by varying the frequency of the carrier signal according to the information signal. Unlike AM, where the amplitude changes, FM keeps the amplitude constant while altering the carrier's frequency. The frequency deviation corresponds to the audio signal's instantaneous amplitude. An FM signal consists of a carrier and frequency deviations that encode the information. One of FM's primary advantages is its resilience to amplitude variations, making it less susceptible to noise and interference. This results in clearer and more reliable communication, particularly in environments with high levels of electromagnetic noise. However, FM requires more complex circuitry and a wider bandwidth than AM.

Generating an FM signal involves using a Voltage-Controlled Oscillator (VCO) in the transmitter. The VCO's audio signal modulates the VCO's frequency, resulting in a frequency-modulated carrier. The modulated signal then passes through an amplifier to increase its strength before transmission. On the receiving end, FM receivers use demodulator circuits to extract the audio signal from the frequency-modulated carrier, which locks onto the carrier's frequency and produces a voltage corresponding to the frequency deviations. This voltage is then converted back into the original audio signal.

FM is the process that changes the instantaneous frequency of a radio frequency wave to convey information. In FM, the frequency of the carrier wave is varied in response to the signal being transmitted. For example, when transmitting voice or music, the frequency of the carrier wave changes in sync with the audio signal's amplitude. This allows the signal to be encoded and then decoded by the receiver to reproduce the original sound or data.

What is the name of the process that changes the instantaneous frequency of an RF wave to convey information? (G8A03)

A. Frequency convolution
B. Frequency transformation
C. Frequency conversion
D. Frequency modulation

Both AM and FM have unique advantages and drawbacks for ham radio. Understanding the differences and principles enhances your ability to choose the appropriate modulation technique for your communication needs.

Phase Modulation

Phase Modulation (PM) is a type of modulation used in radio communication where the phase angle of a carrier signal is changed to encode information. In simple terms, the phase of the carrier wave is shifted in response to the amplitude of the input signal, which could be voice or data. This modulation technique is advantageous in certain types of digital communication and is closely related to frequency modulation. However, the two operate on different principles.

Phase Modulation

Amplitude   Message Signal

Time

Amplitude   Carrier Signal

Time

Amplitude   Phase Modulated Signal

Time

Phase modulation (PM) is the process of changing the phase angle of a radio frequency signal to convey information. The phase angle refers to the position of the waveform relative to time. This angle is adjusted in phase modulation according to the input signal. By shifting the phase of the carrier signal in proportion to the input, the information is encoded and can be transmitted over the air.

What is the name of the process that changes the phase angle of an RF signal to convey information? (G8A02)

A. Phase convolution
B. Phase modulation
C. Phase transformation
D. Phase inversion

Phase modulation is the "shifting of the angle" of the carrier wave to carry data. This differs from amplitude modulation, where the signal's amplitude changes, and from frequency modulation, where the frequency varies. PM is all about how the signal's phase shifts to encode information.

Reactance Modulators

A reactance modulator is a specific device used with a transmitter's RF amplifier stage to produce phase modulation. Reactance modulators work by varying the reactance (which can be inductive or capacitive) in a circuit, thereby shifting the phase of the carrier signal.

When a reactance modulator is connected to a transmitter's RF amplifier stage, it produces phase modulation. The reactance modulator adjusts the carrier wave's phase by changing the circuit's reactive components (inductance or capacitance). This results in the RF signal's varied phase following the input signal, which conveys the transmitted information. Phase modulation alters the timing of the wave, unlike amplitude modulation, which changes the strength of the wave.

What emission is produced by a reactance modulator connected to a transmitter RF amplifier stage? (G8A04)

A. Multiplex modulation
B. Phase modulation
C. Amplitude modulation
D. Pulse modulation

To remember this for the exam, associate a reactance modulator with phase shifts—it's a way of encoding information by adjusting the phase of the signal.

Digital Modulation

Digital modulation has transformed ham radio by offering significant advantages over traditional analog techniques. Instead of varying continuous signals as analog modulation does, digital modulation encodes information into discrete signals. This approach enhances the signal-to-noise ratio, resulting in clearer and more reliable transmissions, even in challenging conditions. It also maximizes bandwidth efficiency, enabling more data to be transmitted over narrower frequency ranges compared to the broader bandwidth requirements of analog modulation, which is more prone to noise and interference.

While the variety of digital modes may initially seem daunting, understanding their underlying modulation techniques—such as frequency shift, phase shift, or amplitude shift—helps operators choose the best mode for their specific needs. Each digital mode has a distinct purpose, whether it's long-distance communication under weak-signal conditions, efficient data transmission in noisy environments, or voice clarity over local repeaters.

By organizing digital modes by modulation type and use case, this section simplifies the learning process and prepares operators to adapt to diverse operating scenarios. Whether you're working DX with FT8, engaging in low-bandwidth text exchanges with PSK31, or using DMR for local digital voice communication, you'll gain the insights needed to match the right mode to the right purpose.

Types of Digital Modulation

Here's an organized list of digital modulation techniques, their modes, and when a ham radio operator would typically use each type:

<u>Amplitude Shift Keying (ASK)</u>

- **On-Off Keying (OOK)**

  - **Use Case:** Rarely used in modern ham radio but foundational for understanding Morse code. OOK was one of the earliest methods for transmitting data via simple on/off carrier pulses.

  - **When to Use:** Historical experiments or educational demonstrations of early communication methods.

<u>Frequency Shift Keying (FSK)</u>

- **RTTY (Radio Teletype)**

  - **Use Case:** Reliable text communication over long distances, often on HF bands.

  - **When to Use:** When you need a robust text-based mode with moderate bandwidth, use in contests or routine communication.

- **AFSK (Audio Frequency Shift Keying)**

  - **Use Case:** Used in APRS (Automatic Packet Reporting System) and some packet radio for real-time data transmission.

  - **When to Use:** For tracking, weather reporting, and sharing text messages or small files over VHF/UHF.

- **FT8 and FT4**

  - **Use Case:** Ultra-efficient weak-signal communication.

  - **When to Use:** When conditions are poor (low sunspot activity, high noise) or when working DX stations with minimal power.

- **Packet Radio (AX.25)**

  - **Use Case:** Reliable data transfer and file sharing, often on VHF/UHF.

  - **When to Use:** For digital communication in local ham networks, message relays, and emergency communications.

- **JT65/JT9**

  - **Use Case:** Ultra-weak signal communication with precise timing and narrow bandwidth.

  - **When to Use:** For very long-distance communication, often on HF or when atmospheric noise levels are high.

- **WSPR (Weak Signal Propagation Reporter)**

  - **Use Case:** Propagation analysis to map signal paths globally with very low power.

  - **When to Use:** To test antennas, propagation conditions, or experimental setups.

Phase Shift Keying (PSK)

- **BPSK (Binary Phase Shift Keying)**

  - **Use Case:** Simple phase modulation for low-power and robust communications.

  - **When to Use:** When working with very narrow bandwidth modes or basic digital communication.

- **QPSK (Quadrature Phase Shift Keying)**

  - **Use Case:** Efficient high-speed data transmission.

  - **When to Use:** For advanced digital modes where higher data rates are needed, such as satellite communication.

- **PSK31**

  - **Use Case:** Keyboard-to-keyboard text communication with narrow bandwidth.

  - **When to Use:** For casual ragchews and when you want to operate at low power but still achieve reliable text exchange on HF.

- **MFSK (Multiple Frequency Shift Keying)**

  - **Use Case:** Combines FSK with multiple tones for robustness against noise and interference.

  - **When to Use:** When operating in noisy HF conditions or requiring error-resistant text-based communication.

Quadrature Amplitude Modulation (QAM)

- **QAM (General Use)**

  - **Use Case:** Common in high-data-rate systems like digital TV or satellite links.

  - **When to Use:** Rarely directly used in amateur radio but foundational for SDR and modern communication systems.

- **DVB-S2 (Digital Video Broadcasting Satellite)**

  - **Use Case:** For high-bandwidth applications like amateur TV (ATV).

  - **When to Use:** When transmitting or receiving video or multimedia via satellite or terrestrial amateur radio TV systems.

Hybrid or Other

- **Olivia**

  - **Use Case:** Highly resilient text-based mode that works well in weak and noisy conditions.

  - **When to Use:** When other modes fail due to noise or weak signals, especially for reliable HF communication.

- **Pactor**

  - **Use Case:** Fast and efficient data transfer over HF.

  - **When to Use:** For email or file transfers, often in Winlink systems during emergencies or maritime operations.

- **Winlink**

  - **Use Case:** A global email network that uses various digital modes like Pactor.

  - **When to Use:** When sending or receiving emails from remote locations without internet access.

- **Digital Voice Modes (D-STAR, DMR, Fusion)**

  - **Use Case:** Clear digital voice communication with additional data capabilities.

  - **When to Use:** For local or regional voice communication with linked repeaters or during events and emergencies.

- **FSQ (Fast Simple QSO)**

  - **Use Case:** Fast text messaging between stations, including selective calling.

  - **When to Use:** When efficiency and rapid communication are required, such as during nets or emergencies.

Frequency Shift Keying

Frequency Shift Keying (FSK) is a digital modulation technique where the frequency of the carrier signal is changed to represent digital information, such as binary data.

In binary FSK (BFSK), two distinct frequencies are used to represent binary "1" and "0." This method is widely used in data communication systems because it is simple to implement and provides reliable communication, even in noisy environments. FSK's ability to shift frequencies to convey data makes it a robust and efficient method for transmitting digital signals.

The two distinct frequencies are identified as mark and space in an FSK signal. The mark frequency represents a binary "1" and the space frequency represents a binary "0." The FSK signal can transmit digital data by shifting between these two frequencies. The concept of "mark" and "space" comes from the early days of telegraphy, where the terms were used to describe the presence and absence of a signal.

How are the two separate frequencies of a Frequency Shift Keyed (FSK) signal identified? (G8C11)

A. Dot and dash
B. On and off
C. High and low
D. Mark and space

Direct binary FSK modulation is generated by directly changing an oscillator's frequency with a digital control signal. In this method, a digital signal (representing 1s and 0s) controls the frequency of the oscillator. One frequency represents a binary "1," and another describes a binary "0." By switching between these two frequencies, the oscillator encodes the binary data onto the carrier wave, allowing digital information to be transmitted over radio waves.

How is direct binary FSK modulation generated? (G8A01)

A. By keying an FM transmitter with a sub-audible tone
B. By changing an oscillator's frequency directly with a digital control signal
C. By using a transceiver's computer data interface protocol to change frequencies
D. By reconfiguring the CW keying input to act as a tone generator

---

To remember this for the exam, think of binary FSK as "direct control" of an oscillator's frequency. Each binary state (1 or 0) causes the oscillator to shift to a corresponding frequency, making it a straightforward yet effective method of modulation for digital communication systems.

Baudot Code

Baudot code is an early form of character encoding used for transmitting text in telecommunication systems, particularly in radio teletype (RTTY) communication. It was one of the first digital telegraph and early digital communication codes. Baudot is still in use today in some ham radio applications, especially for RTTY, where it allows text to be sent efficiently over radio waves. The Baudot code is relatively simple and is known for its use of 5 bits to represent each character, allowing for a streamlined method of text communication.

The Baudot code is a 5-bit code that includes additional start and stop bits to ensure the receiving system can synchronize with the incoming characters. Since 5 bits can only represent a limited number of characters, Baudot also uses a system of shift codes to switch between letters and figures, allowing more symbols to be represented in the same 5-bit structure. The start and stop bits help the receiver detect the beginning and end of each character, making the system more reliable for long-distance text communication.

Which of the following describes Baudot code? (G8C04)

A. A 7-bit code with start, stop, and parity bits
B. A code using error detection and correction
C. A 5-bit code with additional start and stop bits
D. A code using SELCAL and LISTEN

Radio Teletype (RTTY)

RTTY (Radio Teletype) is a method of sending text over radio waves using a system of digital signals. It was one of the earliest digital modes in ham radio and telecommunications. In RTTY, characters are encoded into a digital format (often using Baudot code) and then transmitted by shifting between two tones, commonly referred to as frequency-shift keying (FSK).

RTTY signals are characterized by a specific baud rate, commonly 45.45 baud, and distinct mark and space frequencies representing the binary states. RTTY is widely used for text-based communication, allowing operators to exchange messages over long distances. Its robustness makes it a preferred mode for emergency messaging, where reliable and quick communication is vital. RTTY's simplicity and effectiveness have ensured its continued popularity.

FT8

FT8, developed for weak-signal communication, showcases the cutting-edge capabilities of digital modulation. FT8 operates using very short transmission cycles of 15 seconds, allowing for rapid exchanges of information. This mode is designed to work under extremely poor propagation conditions, making it possible to achieve contact even when signals are barely above the noise floor. FT8's narrow bandwidth further enhances its effectiveness, concentrating the signal's power and reducing the likelihood of interference. Practical applications of FT8 include DXing, where operators aim to contact distant stations, and low-power communication, making it an excellent choice for operators with limited transmission power.

Here is a fun, irrelevant fact about FT8: the man who developed FT8 is a Nobel Prize-winning physicist, Joe Taylor. But don't worry, that's not on the exam.

FT8 uses 8-tone frequency shift keying (8-FSK) as its modulation method. In FT8, there are eight possible frequencies (or tones), each representing a different symbol, allowing the transmission of small data packets quickly and efficiently. This method makes FT8 highly effective at transmitting under weak-signal conditions, as the use of multiple tones enables reliable decoding even when signals are faint or compromised by noise.

What type of modulation is used by FT8? (G8A09)

A. 8-tone frequency shift keying
B. Vestigial sideband
C. Amplitude compressed AM
D. 8-bit direct sequence spread spectrum

FT8 is capable of receiving signals with very low signal-to-noise ratios (SNR), making it one of the best narrowband digital modes for weak-signal communication. The mode can decode signals with SNRs as low as -24 dB, meaning it can pull a usable signal out of background noise where many other modes would fail. This makes FT8 especially valuable in crowded or noisy bands or during poor propagation conditions, where other modes might not perform as well.

Which of the following narrowband digital modes can receive signals with very low signal to noise ratios? (G8C07)

A. MSK144
B. FT8
C. AMTOR
D. MFSK32

To remember this for the exam, think of FT8 as a mode that "hears whispers" in the noise. It excels at pulling out weak signals, making it the go-to digital mode when signal strength is very low, but communication is still needed.

FT8 Signal Reports

In FT8, signal reports are expressed in decibels (dB) relative to the noise floor. This signal-to-noise ratio (SNR) indicates how well the signal stands out from the background noise. FT8 is designed to decode signals with very low SNRs, making it a powerful tool for long-distance communication, even under poor conditions.

An FT8 signal report of +3 dB means that the signal received is 3 dB above the noise floor in a 2.5 kHz bandwidth. In simpler terms, the signal is stronger than the background noise by a factor of +3 dB, making it relatively easy to decode. Positive signal reports like +3 indicate a strong signal relative to the noise, while negative values would indicate weaker signals.

What does an FT8 signal report of +3 mean? (G8C15)

A. The signal is 3 times the noise level of an equivalent SSB signal
B. The signal is S3 (weak signals)
C. The signal-to-noise ratio is equivalent to +3dB in a 2.5 kHz bandwidth
D. The signal is 3 dB over S9

And, as a reminder, the 2.5 kHz bandwidth is specified in this question because it represents the typical bandwidth used for receiving digital modes like FT8. Signal-to-noise ratio (SNR) is often measured relative to the bandwidth being used, and in FT8, this is standardized at 2.5 kHz.

Packet Radio

Packet radio is a digital communication mode used to send and receive data over radio waves, similar to how data is transmitted over the internet. In packet radio, data is broken into small units called packets, which are sent individually, allowing for error-checking and reliable communication over long distances. This system enables the transfer of text, files, and even email-like messages between operators, often using VHF or UHF frequencies.

Each packet of data includes a header (with routing & handling information) and the actual message or data. Packet radio networks often use a terminal node controller (TNC) to handle the packetizing of data, error detection, and retransmission of lost packets. The system is popular in ham radio because it can send data even in poor conditions. It is used for messaging, telemetry, and connecting to packet bulletin board systems (BBS) or the internet.

What part of a packet radio frame contains the routing and handling information? (G8C03)

A. Directory
B. Preamble
C. Header
D. Trailer

In essence, packet radio turns radio signals into a form of digital communication, making it possible for hams to exchange data much like emails or text messages, but over the airwaves.

Weak Signal Propagation Reporter

WSPR (Weak Signal Propagation Reporter) is a digital mode for low-power transmission. It is widely used by operators to assess HF propagation. WSPR is part of a system that allows operators to send and receive low-power signals to study how radio waves propagate through the atmosphere. The primary goal of WSPR is to gather data on how signals travel over long distances and through different atmospheric conditions, making it an excellent tool for understanding propagation at various times of the day and in different seasons.

WSPR is a digital mode that acts as a low-power beacon to assess HF propagation. Operators use WSPR to send weak signals that can be decoded even under poor conditions. These signals are transmitted at very low power, often just a few watts or less, and are received by other WSPR stations around the world. By tracking which signals are received and where, amateur radio operators can assess how well HF signals are propagating under current conditions. This makes WSPR an invaluable tool for studying propagation without needing high-power equipment.

Which digital mode is used as a low-power beacon for assessing HF propagation? (G8C02)

A. WSPR
B. MFSK16
C. PSK31
D. SSB-SC

WSPR is a "whisper" in the world of radio. This low-power signal travels far and helps assess how well radio waves move through the atmosphere. It's ideal for propagation testing because it works even with weak signals and challenging conditions.

Phase Shift Keying

Phase Shift Keying 31

PSK31, or Phase Shift Keying 31 baud, exemplifies the advantages of digital modulation. It works by varying the phase of the carrier signal to encode data, providing a robust and efficient communication method. PSK31's signal characteristics include a very narrow bandwidth of approximately 31.25 Hz, which allows multiple signals to coexist within the same frequency range without significant interference. The mode supports low-power communication, making it ideal for operators looking to achieve long-distance contact with minimal power. PSK31 also incorporates error correction features, enhancing the reliability of data transmission. This mode is prevalent in digital contests, where efficient use of spectrum and low power requirements make for a challenging contest.

Varicode

One of the key elements of PSK31 is its use of Varicode, a character encoding system specifically designed for PSK31. Unlike fixed-length codes like ASCII, Varicode uses shorter codes for more frequently used characters and longer codes for less common ones, making data transmission faster and more efficient.

Varicode is a code used in PSK31 to send characters. It is a variable-length encoding system, meaning that more ordinary characters, like vowels or spaces, are represented by shorter sequences of bits. In comparison, less common characters are defined by longer sequences. This makes Varicode very efficient, especially in a mode like PSK31, where bandwidth is limited, and efficiency is key. Using Varicode, PSK31 ensures faster data transmission while maintaining clear communication with minimal bandwidth.

Which type of code is used for sending characters in a PSK31 signal? (G8C12)

A. Varicode
B. Viterbi
C. Volumetric
D. Binary

Varicode is a more innovative way to encode text and makes PSK31 fast and efficient for real-time keyboard-to-keyboard digital communication.

In PSK31, using uppercase letters will slow the transmission because they require longer Varicode bit sequences than lowercase letters or other common characters. This difference in encoding makes uppercase transmissions take longer, which can slightly reduce communication efficiency in this mode. The encoding system is designed this way to optimize for speed in typical usage, where lowercase letters dominate.

Which of the following statements is true about PSK31? (G8C08)

A. Upper case letters are sent with more power
B. Upper case letters use longer Varicode bit sequences and thus slow down transmission
C. Error correction is used to ensure accurate message reception
D. Higher power is needed as compared to RTTY for similar error rates

Quadrature Phase Shift Keying

QPSK (Quadrature Phase Shift Keying) is a digital modulation technique used in radio communication. In QPSK, the phase of the carrier signal is changed to represent data, but unlike simpler phase-shift keying methods (such as binary PSK, which only shifts the phase by 180 degrees), QPSK uses four distinct phase shifts: 0°, 90°, 180°, and 270°. This allows QPSK to transmit two bits of data per symbol, effectively doubling the data rate compared to binary PSK.

Said again, QPSK modulation transmits digital data by shifting the phase of the carrier signal in one of four possible angles: 0°, 90°, 180°, or 270°. Each phase shift represents a pair of bits, also known as a "dibit." For example, 0° might represent "00," 90° might represent "01," and so on. Using four distinct phase shifts, QPSK efficiently encodes two bits per symbol, allowing faster data transmission without requiring more bandwidth.

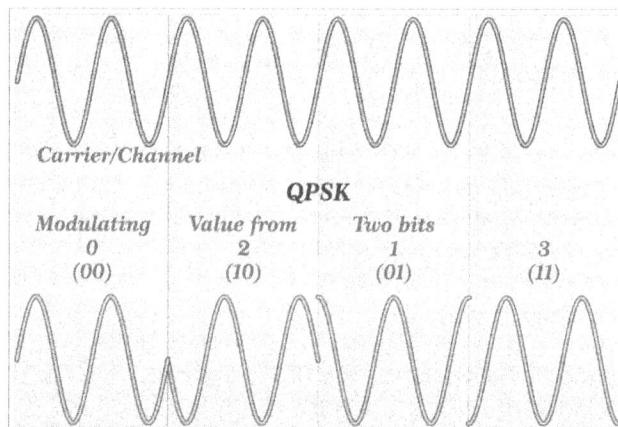

Carrier/Channel

QPSK

| Modulating 0 (00) | Value from 2 (10) | Two bits 1 (01) | 3 (11) |

What is QPSK modulation? (G8A12)

A. Modulation using quasi-parallel to serial conversion to reduce bandwidth
B. Modulation using quadra-pole sideband keying to generate spread spectrum signals
C. Modulation using Fast Fourier Transforms to generate frequencies at the first, second, third, and fourth harmonics of the carrier frequency to improve noise immunity
D. Modulation in which digital data is transmitted using 0-, 90-, 180- and 270-degrees phase shift to represent pairs of bits

QPSK transmits two bits of data with each phase shift, making it more efficient than simpler schemes like BPSK, which uses only two-phase shifts. This efficiency makes QPSK ideal for high-speed systems that require fast data transmission.

By encoding two bits per symbol, QPSK improves bandwidth efficiency while maintaining robustness against noise. It is commonly used in satellite communications, wireless networks, and digital TV for its ability to handle larger data loads.

QPSK31

QPSK31 is a digital mode used in ham radio communication that combines Quadrature Phase Shift Keying with a baud rate of 31.25 (hence the "31" in the name). QPSK31 is similar to PSK31, but it offers greater robustness due to its error correction capabilities. Like PSK31, it is narrowband, making it very efficient in terms of bandwidth use. QPSK31's encoding method allows for reliable communication, even under poor signal conditions, and it is particularly suitable for low-power, long-distance communication, commonly known as "weak signal" work.

QPSK31 characteristics:

- Sideband Sensitive: QPSK31 is sensitive to the sideband (upper or lower) in which it is transmitted. If the sideband is incorrect, the signal might not decode properly, making it crucial to use the correct sideband for transmission.

- Error Correction: QPSK31 incorporates error correction coding, which makes it more resilient to interference and signal degradation. This means that the system can automatically detect and correct some transmission errors, improving the reliability of the communication.

- Narrow Bandwidth: QPSK31 has a very narrow bandwidth, approximately the same as BPSK31 (Binary Phase Shift Keying). Both modes use around 31.25 Hz of bandwidth, making them ideal for communicating over long distances using low power.

Which of the following is characteristic of QPSK31? (G8A06)

A. It is sideband sensitive
B. Its encoding provides error correction
C. Its bandwidth is approximately the same as BPSK31
D. All these choices are correct

Focus on the fact that QPSK31 is sideband sensitive, uses error correction to improve signal reliability, and operates within the same narrow bandwidth as BPSK31, which helps with efficient, long-distance communication.

But wait... what's BPSK31?

Binary Phase Shift Keying

BPSK31 stands for Binary Phase Shift Keying 31.25 baud, and it is a popular digital mode used in ham radio for text-based communication. It operates by modulating the phase of a carrier signal between two distinct states (0° and 180°), which represents binary data (1s and 0s). The "31" in BPSK31 refers to the baud rate, which is 31.25 baud, meaning it transmits 31.25 symbols per second, making it an efficient mode for weak signal and low-power communication.

BPSK31 is well known for its narrow bandwidth of approximately 31 Hz. Since it uses minimal bandwidth, it can work effectively even when signal strength is low, and it is often used for keyboard-to-keyboard digital communication. Operators appreciate BPSK31's simplicity and ability to maintain communication when voice modes fail due to poor signal conditions.

Overmodulation

Overmodulation occurs when the amplitude of the modulating signal exceeds the limits that the transmitter or system is designed to handle. In simpler terms, it's when the input signal is too strong for the transmitter, leading to distortion and interference in the transmitted signal. This often causes the signal to "spill over" into adjacent frequencies, creating unwanted side effects like distortion and interference with other communications.

In amplitude modulation (AM), for example, overmodulation can cause the peaks of the modulated signal to be "clipped," which leads to signal distortion. In frequency modulation (FM) or digital modes, overmodulation results in a signal that occupies more bandwidth than intended, causing interference and degraded transmission quality.

Preventing overmodulation is crucial to ensure a clear, high-quality signal and to avoid interfering with nearby frequencies or other operators. It can often be managed by reducing the input signal level to the transmitter.

AM Overmodulation

In amplitude modulation (AM) and other modulation methods, maintaining a proper balance between the input signal and the carrier wave is essential for clear transmission. Suppose the modulation goes beyond the allowable level. In that case, it can lead to signal distortion, poor transmission quality, and unintended interference with other signals. Overmodulation is a common issue that can result in unwanted side effects, such as excessive bandwidth, splatter, or interference with adjacent frequencies.

One of the main effects of overmodulation is the production of excessive bandwidth. When a signal is overmodulated, it causes the transmitter to generate unwanted harmonics and sidebands that spread beyond the intended frequency range. This results in the signal occupying more bandwidth than necessary, potentially causing interference with adjacent channels and other signals.

Which of the following is an effect of overmodulation? (G8A08)

A. Insufficient audio
B. Insufficient bandwidth
C. Frequency drift
D. Excessive bandwidth

This not only degrades the quality of the transmission but also causes interference, which is detrimental to both the operator and others sharing the frequency spectrum.

Digital Overmodulation

Waterfall Displays

A waterfall display visualizes signals received over time in a particular frequency range, commonly used in digital modes and radio teletype (RTTY). On this display, time flows downward while frequency is shown across the horizontal axis, and the intensity or strength of the signal is displayed using color gradients. This tool helps operators visualize multiple signals, their strength, and any interference in real-time. Understanding how to interpret the patterns on a waterfall display is key for identifying issues like overmodulation and adjusting your transmission settings.

On a waterfall display, overmodulation appears as one or more vertical lines on either side of the data mode or RTTY signal. Overmodulation occurs when the input level to the transmitter is too high, causing distortion and unwanted side effects like these vertical lines. This can interfere with other signals and reduce the clarity of your own transmission. The vertical lines indicate that the signal has exceeded the intended modulation limits, spilling into adjacent frequencies, which can cause interference.

What is indicated on a waterfall display by one or more vertical lines on either side of a data mode or RTTY signal? (G8C13)

A. Long path propagation
B. Backscatter propagation
C. Insufficient modulation
D. Overmodulation

Frequency is represented along the horizontal axis, showing a range of frequencies, you are monitoring. Signal strength is indicated by the intensity or color brightness; stronger signals appear more intense or brighter. Time is represented vertically, meaning older signals are shown at the top, and more recent signals are displayed at the bottom as the display updates in real-time.

Which of the following describes a waterfall display? (G8C14)

A. Frequency is horizontal, signal strength is vertical, time is intensity
B. Frequency is vertical, signal strength is intensity, time is horizontal
C. Frequency is horizontal, signal strength is intensity, time is vertical
D. Frequency is vertical, signal strength is horizontal, time is intensity

---

Remember overmodulation as "spilling over" onto other frequencies, and the vertical lines on the waterfall display are a visual sign that the signal is too strong and needs to be adjusted to prevent interference.

Unmodulated Carrier

An unmodulated carrier is a continuous radio frequency signal transmitted without any additional information or modulation. In radio communication, the carrier wave serves as the base signal that can be modulated by varying its amplitude, frequency, or phase to carry voice, data, or other information. However, when the carrier is unmodulated, it remains constant in amplitude and frequency, and no information is transmitted.

An unmodulated carrier is often used as a reference signal or during testing and calibration of transmitters and receivers. It doesn't carry any meaningful communication, but it provides a foundation for modulation to be added later.

What is the ratio of PEP to average power for an unmodulated carrier? (G5B11)

A. 0.707
B. 1.00
C. 1.414
D. 2.00

The ratio of Peak Envelope Power (PEP) to average power for an unmodulated carrier is 1.00. This means that the PEP and the average power are the same for an unmodulated carrier. Since an unmodulated carrier is a steady, continuous signal with no variations in amplitude, there are no peaks in power beyond the average level, so the PEP is equal to the average power.

To remember this for the exam, think of an unmodulated carrier as a signal without fluctuations. Since PEP measures the maximum power in a signal's envelope, and an unmodulated carrier has no variations, the maximum power (PEP) and the average power are identical, giving a 1:1 ratio.

Try this one:

What is the output PEP of an unmodulated carrier if the average power is 1060 watts? (G5B13)

A. 530 watts
B. 1060 watts
C. 1500 watts
D. 2120 watts

An unmodulated carrier's output PEP equals its average power, a 1 to 1 ratio. In the case of an unmodulated carrier, there are no variations in the signal, so the average power and PEP are the same. Therefore, if the average power is 1060 watts, the PEP is also 1060 watts.

Remember that there are no peaks or dips in power for an unmodulated carrier, as it is a steady signal. Since PEP measures the maximum power of a signal's envelope, and an unmodulated carrier is constant, its PEP is the same as its average power.

Data Reliability and Networking

In the world of digital communication, transmitting data accurately and efficiently is as critical as sending the signal itself. While modulation techniques lay the foundation for encoding and transmitting information, maintaining data integrity and managing networks are the keys to seamless communication.

This section explores the tools and strategies that ensure reliable data exchange, even in challenging conditions. From Automatic Repeat reQuest (ARQ), which handles retransmissions to correct errors, to Forward Error Correction (FEC), which minimizes disruptions by proactively encoding redundancy, we'll dive into how these systems work together to enhance communication.

Additionally, we'll discuss mesh networks, an innovative approach to creating resilient and adaptive connections, especially in emergency or remote environments. By understanding these concepts, you'll gain a comprehensive view of how modern radio systems achieve reliability and adaptability, empowering you to operate more effectively in any scenario.

Automatic Repeat reQuest

ARQ (Automatic Repeat reQuest) is a communication protocol used in digital radio and data transmission systems to ensure the accurate delivery of data. In ARQ mode, the transmitter sends a packet of data and then waits for an acknowledgment from the receiver. If the receiver detects any errors in the data, it sends a request back to the transmitter to resend the data. This process continues until the data is correctly received and acknowledged, ensuring reliable communication even in noisy or weak signal conditions.

Key Features of ARQ

- Error detection: The receiver checks for errors in the transmitted data.

- Automatic retransmission: If errors are detected, the receiver requests a retransmission.

- Reliability: ARQ ensures that the data is accurately transmitted by repeating the process until the data is error-free.

ARQ is often used in systems where data accuracy is critical, such as in digital communication modes and internet communication protocols like TCP/IP. In amateur radio, it is used in modes like Pactor and other digital modes requiring error-free communication.

This mode is highly efficient for maintaining the integrity of transmitted data, particularly in situations where interference or weak signals may corrupt the information being sent.

Transmission Attempts

In ARQ mode, the system will drop the connection if there are excessive transmission attempts without successfully exchanging information. This happens because there is a limit to how many times the system will try to resend a packet after receiving a NAK (Negative Acknowledgment) due to errors. If too many attempts fail, the connection is terminated, as it becomes clear that the communication link is unreliable at that moment.

What action results from a failure to exchange information due to excessive transmission attempts when using an ARQ mode? (G8C06)

A. The checksum overflows
B. The connection is dropped
C. Packets will be routed incorrectly
D. Encoding reverts to the default character set

Consider ARQ, a system that knows when to "give up" after too many failed retries. If it can't send data correctly after a set number of tries, the connection is dropped to prevent wasting time and resources on an unstable link.

Negative Acknowledgment

In ARQ mode, a NAK (Negative Acknowledgment) response means that the receiver has detected an error in the transmitted packet and is requesting the transmitter to retransmit the packet. The NAK serves as a signal to the transmitter that the data was not correctly received. This process continues until the receiver gets the correct data and sends an acknowledgment (ACK). By using NAK and ACK responses, ARQ ensures accurate communication by automatically handling errors without manual intervention.

In an ARQ mode, what is meant by a NAK response to a transmitted packet? (G8C05)

A. Request retransmission of the packet
B. Packet was received without error
C. Receiving station connected and ready for transmissions
D. Entire file received correctly

To remember this for the exam, think of NAK as a "No, send it again" response. It's part of the system that ensures reliable data transfer by requesting retransmission when needed.

Forward Error Correction

Forward Error Correction (FEC) is a technique used in digital communication systems to enhance the reliability of data transmission by allowing the receiver to detect and correct errors without needing to request a retransmission. In FEC, redundant information is added to the original data before transmission. This redundancy allows the receiver to identify and fix errors that may occur due to noise, interference, or weak signals. FEC is commonly used in radio communication, satellite links, and digital modes to improve communication in challenging conditions.

FEC works by sending redundant information along with the actual data. For example, suppose a bit of data is lost or corrupted during transmission. In that case, the redundant information provides clues that help the receiver reconstruct the correct data. This method improves communication reliability, especially in environments where retransmissions are not practical, or signals are weak.

How does forward error correction (FEC) allow the receiver to correct data errors? (G8C10)

A. By controlling transmitter output power for optimum signal strength
B. By using the Varicode character set
C. By transmitting redundant information with the data
D. By using a parity bit with each character

FEC is a way of "building in safety nets" with extra data, so even if part of the message is lost or distorted, the receiver can still piece it together correctly without asking for a resend.

Mesh Networks

Mesh networks are a type of communication network where each node, or station, is connected to multiple other nodes, creating a robust and flexible network. These nodes communicate with each other to forward data. If one node fails, the network can dynamically reroute the data through alternate nodes. This makes mesh networks highly reliable, especially when individual nodes become unavailable or damaged. Mesh networks are commonly used in microwave communications. They are becoming more popular in amateur radio for their reliability and ease of setup.

In a mesh network of microwave nodes, if one node fails or becomes unreachable, the network can reroute the data packet through another available node to ensure that the packet still reaches its target. This is possible because each node in a mesh network is connected to multiple other nodes, allowing alternate paths to be used if the primary route is blocked. This redundancy makes mesh networks highly resilient to node failures and other issues that might disrupt communication in more traditional network setups.

Which is true of mesh network microwave nodes? (G8C09)

A. Having more nodes increases signal strengths
B. If one node fails, a packet may still reach its target station via an alternate node
C. Links between two nodes in a network may have different frequencies and bandwidths
D. More nodes reduce overall microwave out of band interference

---

Tools like ARQ and FEC enhance accuracy and minimize errors, while mesh networks offer flexibility and adaptability, especially in complex or challenging environments. Together, these technologies form the backbone of modern digital communication, empowering operators to maintain seamless connections under varying conditions. By mastering these principles, you'll be well-equipped to ensure reliable communication, whether you're exchanging data in everyday operations or tackling the challenges of emergency scenarios.

Bandwidth

Bandwidth refers to the range of frequencies a signal occupies. Different modulation methods use different amounts of bandwidth, which affects how efficiently the frequency spectrum is used and how much interference might occur. Phone emissions, such as AM, FM, and single sideband, vary in bandwidth requirements. For efficient communication, especially in crowded frequency bands, using narrower bandwidth signals is often preferred because it reduces the chance of interference and allows more signals to occupy the same frequency range.

Single Sideband

SSB is a form of amplitude modulation that uses the narrowest bandwidth among phone emissions. In SSB, only one of the two sidebands (either the upper sideband or the lower sideband) is transmitted, and the carrier is either significantly reduced or entirely suppressed. This results in a much narrower signal than traditional AM or FM, making SSB more efficient in spectrum use. SSB typically occupies about 2.5 to 3 kHz of bandwidth, while AM can use up to 6 kHz or more.

Which of the following phone emissions uses the narrowest bandwidth? (G8A07)

A. Single sideband
B. Vestigial sideband
C. Phase modulation
D. Frequency modulation

SSB is the "efficient" modulation mode, using the smallest bandwidth while allowing clear voice communication. This makes it a popular choice in amateur radio and long-distance communication, where minimizing bandwidth is crucial.

Frequency Modulation

In frequency modulation (FM), the total bandwidth of a signal depends on two key factors: the deviation and the modulating frequency. The deviation refers to how much the carrier frequency shifts from its center frequency in response to the input signal. In contrast, the modulating frequency is the highest in the signal being transmitted (like voice or music).

To calculate the total bandwidth of an FM transmission, you use Carson's Rule, which estimates the bandwidth as follows:

$$Total\ Bandwidth = 2 \times (Deviation + Modulating\ Frequency)$$

What is the total bandwidth of an FM phone transmission having 5 kHz deviation and 3 kHz modulating frequency? (G8B06)

A. 3 kHz
B. 5 kHz
C. 8 kHz
D. 16 kHz

Given an FM transmission with a 5 kHz deviation and a 3 kHz modulating frequency, Carson's Rule gives us the following:

$$Total\ Bandwidth = 2 \times (5\ kHz + 3\ kHz) = 2 \times 8\ kHz = 16\ kHz$$

Thus, the total bandwidth of the FM transmission is 16 kHz.

Carson's Rule helps you estimate the bandwidth of an FM signal by doubling the sum of the deviation and the modulating frequency. This formula gives you a quick way to figure out the approximate bandwidth an FM signal uses.

Symbol Rate and Bandwidth

Symbol rate refers to how many symbols are transmitted per second in a digital communication system. Each symbol represents a unit of data, which can include one or more bits. The bandwidth is the range of frequencies required to transmit the signal without distortion. Understanding the relationship between symbol rate and bandwidth is crucial in radio communication because the faster you send data (higher symbol rate), the more bandwidth is required to transmit it effectively.

A direct relationship exists between transmitted symbol rate and bandwidth: higher symbol rates require wider bandwidth. When you transmit more symbols per second, you are packing more information into the signal, increasing the frequency range needed to carry that information. The transmitted data will become distorted or lost if the bandwidth is too narrow for the symbol rate. The bandwidth must be wide enough to accommodate the higher symbol rate to ensure clear, efficient communication.

What is the relationship between transmitted symbol rate and bandwidth? (G8B10)

A. Symbol rate and bandwidth are not related
B. Higher symbol rates require wider bandwidth
C. Lower symbol rates require wider bandwidth
D. Bandwidth is half the symbol rate

Consider symbol rate as how fast you are "talking" in data terms. The faster you "talk" (higher symbol rate), the more room (wider bandwidth) you need for the signal to pass through clearly.

Chapter Summary

In summary, modulation and bandwidth are fundamental aspects of radio communication, influencing signal clarity and transmission range. Modulation enables signals to convey voice, data, or video by modifying a carrier wave with information. Bandwidth, on the other hand, determines the spectrum space that modulated signals occupy and affects the ability to differentiate multiple signals within the same frequency range.

Understanding how modulation types impact bandwidth helps radio operators select appropriate methods for specific applications, balancing clarity, efficiency, and signal range to optimize communication.

Chapter 6

Frequency

RADIO WAVE PROPAGATION AND frequency are closely related because the frequency of a radio wave determines how it interacts with the atmosphere and the ionosphere. Lower frequencies, like those in the HF range, can be refracted or bent back to Earth by the ionosphere, enabling long-distance communication.

However, this ability depends on factors like solar activity and the state of the ionosphere at the time. Higher frequencies, like VHF and UHF, typically pass through the ionosphere and are not reflected, which limits them to line-of-sight communication.

Low vs High HF Bands

Time for a definition review. You've probably seen this idea already in this book: low high-frequency and high high-frequency. So, what are they?

The terms lower high-frequency (HF) bands and higher high-frequency (HF) bands refer to specific segments within the HF spectrum, which spans from 3 MHz to 30 MHz. The HF spectrum is widely used in amateur radio for long-distance communication due to its ability to reflect off the ionosphere.

- **Lower HF Bands**: Typically refers to frequencies closer to the **3 MHz to 10 MHz** range. These include bands like **80 meters (3.5–4.0 MHz)** and **40 meters (7.0–7.3 MHz)**. These frequencies:

  ○ Are more reliable for nighttime communication due to better ionospheric conditions.

  ○ Are more prone to ground wave propagation, making them useful for shorter-range communication during the day.

  ○ Tend to have higher atmospheric noise levels, such as static from thunderstorms.

- **Higher HF Bands**: Typically refers to frequencies in the **10 MHz to 30 MHz** range, including bands like **20 meters (14.0–14.35 MHz)**, **15 meters (21.0–21.45 MHz)**, and **10 meters (28.0–29.7 MHz)**. These frequencies:

  ○ Are more effective for daytime ionospheric propagation, especially during high solar activity.

  ○ Are better suited for long-distance communication (DX) as they reflect well off the ionosphere.

  ○ Generally, experience lower atmospheric noise, leading to cleaner signals.

In Summary:

- **Lower HF bands**: Better at night, shorter-range, more noise.

- **Higher HF bands**: Better during the day, long-distance, less noise.

Understanding how these bands behave helps operators choose the right frequency for optimal communication based on time, conditions, and desired range.

Intermediate Frequency (IF)

Intermediate Frequency (IF) is a crucial concept in superheterodyne receivers, which are widely used in radio communication. A superheterodyne receiver aims to convert the incoming radio signal to a lower, fixed frequency—called the intermediate frequency—before the signal is processed further. This conversion makes it easier to filter, amplify, and demodulate the signal because working with a fixed frequency simplifies the design and improves the performance of the receiver.

Why Use an Intermediate Frequency?

When a radio receives signals from different frequencies, directly amplifying and processing those signals can be difficult, especially at very high frequencies. By mixing the incoming signal with a signal generated by the local oscillator, the receiver creates new frequencies: the sum and difference of the incoming signal and the oscillator's frequency. The difference frequency, which is lower than the original, becomes the intermediate frequency (IF).

Once the signal is converted to IF, the radio can use highly effective filters and amplifiers explicitly designed for that frequency. This makes the receiver more selective and sensitive, improving its ability to pick out weak signals and reject interference from nearby frequencies.

Example of How IF Works:

If a receiver is designed to work with an intermediate frequency of 455 kHz, and it is tuned to receive a signal at 10 MHz, the local oscillator might be set to 10.455 MHz. When the 10 MHz signal mixes with the 10.455 MHz signal from the local oscillator, it creates a difference frequency of 455 kHz, which is the intermediate frequency that the receiver will process.

Key Points to Remember:

- Intermediate frequency simplifies signal processing by converting high-frequency signals to a lower, fixed frequency.

- IF allows for better filtering, amplification, and tuning in a radio receiver, making the radio more effective at isolating and processing the desired signal.

- Superheterodyne receivers, which use IF, are the most common type of radio receivers due to their superior performance.

Signal Processing and Frequency Control

Mixers & Local Oscillators

In a radio receiver, particularly a superheterodyne receiver, the process of converting incoming radio signals to a lower, more manageable frequency called the intermediate frequency (IF) is necessary for efficient signal processing. This conversion happens through a mixer component, which combines the incoming radio signal and the local oscillator (LO) signal. By adjusting or tuning the local oscillator, the receiver can convert a wide range of incoming frequencies to the same intermediate frequency, making it easier to filter, amplify, and demodulate the desired signal.

Which mixer input is varied or tuned to convert signals of different frequencies to an intermediate frequency (IF)? (G8B01)

A. Image frequency
B. Local oscillator
C. RF input
D. Beat frequency oscillator

The local oscillator is the part of the mixer circuit that is varied or tuned to convert signals of different frequencies to a fixed intermediate frequency (IF). The local oscillator generates a signal that mixes with the incoming radio signal in the mixer. This mixing results in two new frequencies: the sum and difference of the local oscillator frequency and the incoming signal frequency. The receiver then selects the difference frequency, which is the intermediate frequency, for further processing.

Mixers and Frequency Conversion

A mixer is a crucial component in radio receivers, and it is used to shift the frequency of incoming radio signals to a lower intermediate frequency (IF). The mixer combines the incoming signal (RF input) with a signal generated by the Local Oscillator (LO). This process creates two new frequencies: one at the sum of the two input frequencies and another at the difference. These new frequencies are key because they allow the receiver to convert a wide range of signals to a fixed intermediate frequency for easier processing.

What combination of a mixer's Local Oscillator (LO) and RF input frequencies is found in the output? (G8B11)

A. The ratio
B. The average
C. The sum and difference
D. The arithmetic product

When an RF signal enters a mixer, it is combined with the frequency generated by the Local Oscillator (LO). The output of the mixer includes two new frequencies: the sum and the difference of the LO frequency and the RF input frequency. For example, if the RF input is 10 MHz and the LO is set to 11 MHz, the mixer output will include frequencies at 21 MHz (sum of 11 MHz + 10 MHz) and 1 MHz (difference of 11 MHz – 10 MHz).

The receiver typically selects the difference frequency (the intermediate frequency, or IF) for further processing, as it's easier to handle a lower frequency in terms of filtering and amplification.

Heterodyning

Let's use some more repetition here to help this concept stick. Heterodyning is when two signals of different frequencies are mixed to produce new frequencies called sum and difference frequencies. Said another way, heterodyning is the mixing of two RF signals. The result of this mixing can generate a signal that is easier to process or decode, especially in receivers. Heterodyning is the core principle behind how most radio receivers work.

What is another term for the mixing of two RF signals? (G8B03)

A. Heterodyning
B. Synthesizing
C. Frequency inversion
D. Phase inversion

In a superheterodyne receiver, the incoming radio frequency (RF) signal is mixed with a locally generated signal (called the local oscillator, or LO). This mixing creates two new frequencies: one at the sum of the two signals and one at the difference. The receiver then selects the difference frequency, called the intermediate frequency (IF), which is lower and easier to amplify and process than the original high-frequency signal.

Heterodyning allows radios to operate more efficiently by shifting high-frequency signals to lower, more manageable frequencies for further processing. This process is vital in many radio technologies, from ham radio receivers to commercial AM and FM radio, as it improves the ability to filter, amplify, and demodulate signals.

Frequency Multipliers

A frequency multiplier is a circuit or device that increases the frequency of an input signal by generating harmonics of that signal. These harmonics are multiples of the original frequency, such as twice, three times, or four times the input frequency. The multiplier circuit selects the desired harmonic and uses it to generate a higher-frequency output signal.

For example, if a signal at 10 MHz is fed into a frequency multiplier set to triple the frequency, the output would be 30 MHz. Multipliers are commonly used in radio transmitters, especially in VHF and UHF systems, to efficiently produce higher frequencies from lower-frequency oscillators, which are easier to control and stabilize.

Key Points to Remember:

- **Multiplies frequency**: Takes an input signal and produces a higher-frequency output (a harmonic multiple).

- **Common in radio transmitters**: Especially useful in VHF/UHF transmitters to generate higher operating frequencies.

VHF FM transmitters

In VHF FM transmitters, sometimes it's easier to generate a lower-frequency signal and then multiply that frequency to reach the desired operating frequency. This is where a frequency multiplier comes into play. A multiplier is a circuit that takes a lower-frequency signal and generates a harmonic (or multiple) of that signal. This method allows the transmitter to efficiently reach higher operating frequencies without directly generating those higher frequencies, which would require more complex and precise oscillators.

In a VHF FM transmitter, the stage that generates a harmonic of a lower frequency signal to reach the desired operating frequency is called a multiplier. For example, suppose the transmitter needs to operate at 150 MHz. In that case, it might start with a signal at 50 MHz and use a frequency multiplier to produce the third harmonic (3 × 50 MHz = 150 MHz). This allows the transmitter to generate high-frequency signals more easily by initially working with lower, more manageable frequencies.

What is the stage in a VHF FM transmitter that generates a harmonic of a lower frequency signal to reach the desired operating frequency? (G8B04)

A. Mixer
B. Reactance modulator
C. Balanced converter
D. Multiplier

A multiplier is a "frequency booster" in a VHF FM transmitter. It takes the original lower frequency and creates a harmonic, or multiple, to bring the signal to the higher operating frequency required for VHF communication.

Frequency Deviation

Frequency deviation in an FM transmitter refers to how much the frequency of the carrier wave shifts from its central frequency in response to the modulation signal. This deviation is directly related to the strength of the input signal (such as voice or data) and is an important parameter because it affects both the quality of the transmission, and the bandwidth used. In radio communication, controlling the deviation is necessary to ensure that the transmitted signal stays within the legal limits and provides clear communication without interference.

What is the frequency deviation for a 12.21 MHz reactance modulated oscillator in a 5 kHz deviation, 146.52 MHz FM phone transmitter? (G8B07)

A. 101.75 Hz
B. 416.7 Hz
C. 5 kHz
D. 60 kHz

First, some key terms in this question.

Key terms:

- **Frequency deviation:** The amount the carrier frequency shifts above or below its center frequency during modulation.

- **Reactance modulated oscillator:** A stage in the transmitter where frequency modulation begins by varying the reactance.

- **Frequency multiplication:** The process of using frequency multipliers in the transmitter to achieve the desired operating frequency (in this case, 146.52 MHz).

FM transmitters often use lower-frequency oscillators modulated with the audio signal, then multiply the frequency to reach the final transmission frequency. During multiplication, the frequency deviation also scales by the same factor as the frequency.

Key Data:

- **Reactance modulated oscillator frequency (initial frequency):** 12.21 MHz

- **Final carrier frequency:** 146.52 MHz

- **Final frequency deviation:** 5 kHz

We must determine the deviation at the **reactance modulated oscillator** before the multiplication occurs.

Step 1: Determine the multiplication factor

The multiplication factor tells us how many times the reactance modulated oscillator's frequency is multiplied to reach the final carrier frequency.

$$Multiplication\ Factor = \frac{Final\ Frequency}{Oscillator\ Frequency}$$

$$Multiplication\ Factor = \frac{146.52\ MHz}{12.21\ MHz} = 12$$

Step 2: Relating deviation to multiplication

The frequency deviation scales by the same factor as the frequency multiplication. Thus:

$$Deviation\ at\ Oscillator = \frac{Final\ Deviation}{Multiplication\ Factor}$$

$$Deviation\ at\ Oscillator = \frac{5\ kHz}{12} = 0.4167\ kHz\ or\ 416.7\ Hz$$

Final Answer

The frequency deviation at the 12.21 MHz reactance modulated oscillator is **416.7 Hz.**

By following this process, you can confidently solve similar questions and understand the interplay between frequency, deviation, and multiplication in FM transmitters.

Interference Types

Image Response

In radio receivers, particularly superheterodyne receivers, image response is an important concept to understand. This occurs when a signal at a different frequency than the desired one gets processed by the receiver, leading to interference. Specifically, image response happens when a signal at twice the intermediate frequency (IF) away from the desired signal frequency is mixed in the receiver. Understanding image response and how to minimize it is critical to ensure clear reception and avoid interference from unwanted signals, especially in crowded frequency environments.

What is the term for interference from a signal at twice the IF frequency from the desired signal? (G8B02)

A. Quadrature response
B. Image response
C. Mixer interference
D. Intermediate interference

An image response is an interference caused by a signal twice the intermediate frequency (IF) away from the desired signal. In a superheterodyne receiver, the local oscillator (LO) generates a frequency that mixes with the incoming radio signal to produce the IF. However, suppose there is a signal at a frequency offset by twice the IF from the desired frequency. In that case, it can also create the same IF, leading the receiver to mistakenly process this unwanted signal. This results in interference and distorted reception.

To prevent image response interference:

1. Use better RF filtering to reject signals at the image frequency before they reach the mixer.

2. Choose a higher IF, as this increases the separation between the desired and image frequencies, making filtering more effective.

By following these steps, you can systematically solve for image response interference and understand how to address it in receiver design.

Think of image response as a "mirror" problem—where a signal that is not intended to be received gets mixed in, creating interference. Image response can be reduced by using a well-designed receiver with adequate filtering to block these unwanted signals.

Intermodulation

Intermodulation occurs when two or more signals are combined in a non-linear circuit, such as an amplifier or antenna system, and produce unwanted spurious outputs, also called intermodulation products. These products are additional frequencies that are not part of the original signals, and they can cause interference with other nearby communications. Intermodulation distortion is common in crowded environments where multiple transmitters are operating, and it is a significant concern in ensuring clean, interference-free communication in ham radio.

What process combines two signals in a non-linear circuit to produce unwanted spurious outputs? (G8B12)

A. Intermodulation
B. Heterodyning
C. Detection
D. Rolloff

Let's sidestep here to discuss non-linear circuits.

Non-Linear Circuits

A non-linear circuit is one in which the relationship between the input and output signals does not follow a straight-line (linear) pattern, meaning the output is not directly proportional to the input. In other words, changes in the input don't produce proportional changes in the output. This behavior occurs because some components in the circuit, like diodes, transistors, and other active devices, have non-linear characteristics. These components do not consistently obey Ohm's law ($V = IR$) across different voltages and currents.

Non-linear circuits are essential in many applications, especially radio and signal processing. For instance, they are used in oscillators, modulators, amplifiers, and frequency mixers, where creating and processing complex waveforms is necessary. Non-linearities in these circuits allow them to perform tasks like signal amplification and frequency modulation, which are vital for communication systems. However, non-linear circuits can also introduce distortion and harmonics, sometimes requiring careful design considerations to minimize unwanted effects.

Intermodulation happens when two signals mix in a non-linear circuit, creating new frequencies that are combinations of the original signals.

These spurious outputs can appear as sums and differences of the original signals. For example:

$$f1 + f2 \quad \text{or} \quad f1 - f2$$

Intermodulation products are classified by their "order," which refers to the sum of the coefficients in the mixing process. For example:

$$2f1 - f2 \quad \text{or} \quad 3f1$$

These unwanted frequencies can interfere with other communications, causing distortion or making it difficult to receive the intended signal.

Intermodulation is a "mixing" problem—two signals interact in a non-linear way, producing additional, unwanted signals that can cause interference. Proper equipment design and isolation can help reduce intermodulation's effects and improve the clarity of radio communication.

How to Achieve It:

1. **Design for Linearity:** Use high-quality components designed to operate within their linear range. Amplifiers, for instance, should be chosen based on their power-handling capacity to avoid overloading and non-linear behavior.

2. **Implement Filtering:** Use bandpass filters to block out-of-band signals that might otherwise interact and create intermodulation products.

3. **Ensure Proper Isolation:** Incorporate shielding, grounding, and physical separation to reduce the coupling of strong external signals into the circuit.

4. **Reduce Signal Overlap:** Design the system to limit the mixing of multiple strong signals, such as by optimizing antenna placement and avoiding high-power transmitters near sensitive receivers.

By addressing intermodulation at its source through design and isolation, you can significantly improve the reliability and clarity of radio communication, especially in environments with many competing signals.

Odd-Order

Odd-order intermodulation products are the ones that appear closest to the original signal frequencies. For example, third-order products (like 2f1–f2) are typically much closer to the original signals than higher-order products. Odd-order products are more problematic for radio operators because they fall near the frequency you're trying to use, potentially causing interference that can be difficult to filter out.

> **Which intermodulation products are closest to the original signal frequencies? (G8B05)**
>
> A. Second harmonics
> B. Even-order
> **C. Odd-order**
> D. Intercept point

To recall this for the exam, remember that odd-order intermodulation products—such as third-order products—are closer to your operating frequencies and more likely to cause interference. This is why managing intermodulation distortion is important, especially in environments with multiple strong signals.

This topic is still a bit fuzzy to me. Let's try to break it down and simplify it.

When two or more signals mix in a non-linear circuit, they can produce unwanted signals called intermodulation products. These extra signals aren't part of the original ones you want and can interfere with your signals and others nearby.

Odd-order intermodulation products, like third-order ones, are the most troublesome because they show up close to the frequencies you're using. This makes them harder to filter out or avoid. To simplify for the exam: Odd-order products are the ones to watch out for because they're nearest to your important signals and can easily cause interference.

Let's break it down into simpler parts:

Odd-Order Intermodulation

Intermodulation products come in different "orders" based on how they are created. An odd-order product of frequencies F1 and F2 means the new signal is created by an odd-number combination of the original signals. For example, a third-order intermodulation product might look like this:

$$2f1 - f2$$

Which of the following is an odd-order intermodulation product of frequencies F1 and F2? (G8B13)

A. 5F1-3F2
B. 3F1-F2
C. 2F1-F2
D. All these choices are correct

This combination of two signals (f1 and f2) is mixed in the non-linear circuit. The result (like 2f1 - f2) is called a third-order product because the numbers add up to 3.

But I still don't understand the 3.

The "order" of an intermodulation product is determined by how many times the original signals are used in a combination. Each signal contributes a number, and we add them up to get the order of the product.

For example, let's look at **2f1 - f2**:

- 2f1 means we're using the first signal (f1) twice.

- f2 means we're using the second signal (f2) once.

So, you add up how many times you're using the signals:

- 2 (from 2f1) + 1 (from f2) = 3 (an odd number so odd-order).

Since 2f1- f2 adds up to 3, this is a third-order product.

Let's look at the other possible answers.

- A. 5F1-3F2 = 5 + 3 = 8 (even number, so an even order)

- B. 3F1-F2 = 3 + 1 = 4 (even number, so an even order)

Chapter Summary

This chapter explored the fundamental role of frequency in amateur radio, explaining how different bands behave and why selecting the right frequency is essential for effective operation. With an overview of how frequencies interact with the ionosphere, detailing why HF bands enable long-distance communication while VHF and UHF bands primarily support line-of-sight transmission. It highlighted the distinctions between lower and higher HF bands, their advantages, and the best times for use. Key factors like time of day, atmospheric conditions, and solar activity are examined to help operators optimize their radio performance.

We then delved into intermediate frequency (IF) and its critical role in superheterodyne receivers, breaking down how mixers and local oscillators convert signals to more manageable frequencies for improved filtering and reception. Essential concepts such as heterodyning, frequency multipliers, and frequency deviation are introduced to explain how signals are processed and refined for clear transmission.

Finally, this chapter addressed common interference issues, including image response and intermodulation, and provides strategies for minimizing unwanted signals that can disrupt communication. By the end of this chapter, readers will have a solid understanding of frequency behavior, signal processing, and the techniques needed to optimize their radio setup for maximum performance.

Chapter 7

Electrical Components

IMAGE YOURSELF AT YOUR workbench, surrounded by various components, each playing a vital role in bringing your ham radio projects to life. As you prepare to upgrade your license, advancing beyond the basics of electrical principles becomes increasingly important. This chapter goes beyond the fundamental elements of electrical circuits and prepares you for the advanced topics on the General exam.

Components

Understanding how electrical components like capacitors, inductors, and their properties interact is fundamental for optimizing your equipment's performance. These concepts—capacitor, inductance, reactance, and resonance—are profoundly connected and influence how radio circuits handle electrical signals. Together, they determine how your radio can tune into different frequencies and filter out unwanted signals.

Reactance

A capacitor stores and releases electrical energy, while capacitance measures its ability to do so. Capacitors play a crucial role in tuning and signal filtering by allowing alternating current to pass while blocking direct current. On the other hand, an inductor stores energy in a magnetic field and opposes changes in current. The property of inductance determines how much resistance an inductor creates to changes in current flow.

Both capacitors and inductors create reactance, which opposes the flow of AC. Still, they do so in opposite ways—capacitors decrease reactance with higher frequencies, and inductors increase it.

What is reactance? (G5A02)

A. Opposition to the flow of direct current caused by resistance
B. Opposition to the flow of alternating current caused by capacitance or inductance
C. Reinforcement of the flow of direct current caused by resistance
D. Reinforcement of the flow of alternating current caused by capacitance or inductance

---

The opposition to the flow of alternating current in a capacitor is called reactance. Unlike resistance, which affects both direct current and AC, reactance only impacts AC signals. In a capacitor, reactance decreases as the frequency of the AC increases. This means that at lower frequencies, the capacitor provides more opposition to the current flow, while at higher frequencies, the opposition (or reactance) becomes smaller, allowing more current to pass through.

Which of the following is opposition to the flow of alternating current in a capacitor? (G5A04)

A. Conductance
B. Reluctance
C. Reactance
D. Admittance

The opposition to the flow of alternating current (AC) in an inductor is called reactance. Just like capacitors, inductors create reactance, but they do so in a different way. In an inductor, reactance increases as the frequency of the AC rises. This means that the inductor offers more opposition to the current flow at higher frequencies, making it harder for the signal to pass through. Reactance in an inductor is called inductive reactance, and it plays a key role in controlling the flow of AC in circuits, especially in tuning and filtering applications in radios.

Which of the following is opposition to the flow of alternating current in an inductor? (G5A03)

A. Conductance
B. Reluctance
C. Admittance
D. Reactance

The unit used to measure reactance is the ohm (Ω), the same unit used to measure resistance. While resistance applies to both direct current and alternating current, reactance explicitly describes the opposition to the flow of AC in components like capacitors and inductors – repetition that builds memory! Just like resistance limits the flow of current, reactance does the same, but its value changes depending on the frequency of the AC signal.

What unit is used to measure reactance? (G5A09)

A. Farad
B. Ohm
C. Ampere
D. Siemens

To remember this for the exam, keep in mind that although reactance behaves differently from resistance (since it varies with frequency), it is still measured in ohms. Whether it's a capacitor or inductor, their reactance in an AC circuit is quantified using the familiar unit of ohms.

What letter is used to represent reactance? (G5A11)

A. Z
B. X
C. B
D. Y

The letter used to represent reactance is X. And no, you don't need to Tweet about that – sorry, that's what they call a bad ham joke! In electrical formulas and circuit analysis, X is used to denote reactance, distinguishing it from resistance, which is defined by the letter R. Reactance can either be from a capacitor (capacitive reactance) or an inductor (inductive reactance). Xc represents capacitive reactance, and Xi represents inductive reactance. Both types of reactance oppose the flow of alternating current.

Inductor & Capacitor Reaction to AC

That was a lot to take in. Let's review two more questions and use that repetitive memory-building technique.

Inductor

An inductor opposes AC flow through inductive reactance. The key thing to remember is that <u>as the frequency of the AC increases, the inductor's reactance also increases</u>. This means that at higher frequencies, the inductor provides more opposition to the current, limiting how much of the signal can pass through. Inductors are often used in circuits where you want to block or filter out higher-frequency signals.

How does an inductor react to AC? (G5A05)

A. As the frequency of the applied AC increases, the reactance decreases
B. As the amplitude of the applied AC increases, the reactance increases
C. As the amplitude of the applied AC increases, the reactance decreases
D. As the frequency of the applied AC increases, the reactance increases

---

Capacitor

A capacitor, on the other hand, behaves in the opposite way. <u>As the frequency of the AC increases, the capacitive reactance decreases.</u> This means capacitors allow more current to pass through as the frequency rises. A capacitor offers more opposition at lower frequencies, but as the frequency increases, it becomes less of a barrier, letting higher-frequency signals pass more easily.

How does a capacitor react to AC? (G5A06)

A. As the frequency of the applied AC increases, the reactance decreases
B. As the frequency of the applied AC increases, the reactance increases
C. As the amplitude of the applied AC increases, the reactance increases
D. As the amplitude of the applied AC increases, the reactance decreases

---

To Sum it Up

Inductors and capacitors oppose AC's flow, but they do so in opposite ways. <u>As the frequency of the applied AC increases, the reactance of an inductor increases,</u> meaning it resists higher frequencies more.

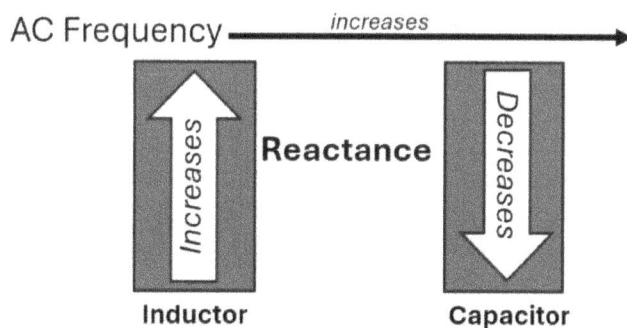

In contrast, <u>a capacitor's reactance decreases as the frequency increases,</u> allowing more current to flow at higher frequencies. This balance between inductors and capacitors is critical in radio circuits, particularly for tuning and filtering different frequencies. Understanding how they react to AC helps you manage which frequencies pass through a circuit and which are blocked.

Impedance

What is impedance? (G5A08)

A. The ratio of current to voltage
B. The product of current and voltage
C. The ratio of voltage to current
D. The product of current and reactance

Impedance is <u>the ratio of voltage to current in an AC circuit.</u> It's similar to resistance in a DC circuit, but impedance considers not only resistance but also the reactance caused by capacitors and inductors in an AC circuit. Impedance is measured in ohms (Ω) and determines how much the circuit resists the flow of alternating current.

Impedance is the overall "opposition" in an AC circuit that combines resistance and reactance. The higher the impedance, the more the circuit resists current flow for a given voltage. The lower the impedance, the easier it is for current to flow through the circuit. This concept is crucial when designing and tuning radio circuits to ensure the right balance between voltage and current for efficient signal transmission and reception.

Admittance

What is the term for the inverse of impedance? (G5A07)

A. Conductance
B. Susceptance
C. Reluctance
D. Admittance

<u>The term for the inverse of impedance is admittance.</u> While impedance measures how much a circuit opposes the flow of AC, admittance measures how easily AC flows through the circuit. In other words, if impedance is the "resistance" to current, admittance is the "permission" for current to flow. Admittance is measured in Siemens (S), the inverse unit of ohms.

Remember, impedance "blocks" current, while admittance "allows" current to pass. The higher the admittance, the easier it is for current to flow through a circuit. This relationship between impedance and admittance is important in understanding how circuits behave, especially in radio and communication systems, where efficient power transfer is needed.

Resonance

When resonance occurs, the reactance of the capacitor and inductor cancels out, allowing the circuit to operate efficiently at a specific frequency. It's the point where the energy exchange between a capacitor and an inductor in an LC circuit is perfectly balanced. When this happens, the circuit can either amplify or filter signals at a specific frequency with maximum efficiency.

Think of it this way: just like a swing moves higher with perfectly timed pushes, a circuit at resonance operates at its best when the alternating current (AC) matches the natural frequency of the LC components. At this resonant frequency, the opposition (reactance) from the capacitor and the inductor cancels each other out, allowing the circuit to pass or amplify the signal with minimal energy loss. This concept is fundamental in radios, where tuning to a particular frequency involves adjusting the circuit to resonate with the signal you want to receive.

What's an LC Circuit?

An LC circuit (also known as a tuned circuit or resonant circuit) is an electrical circuit consisting of an inductor (L) and a capacitor (C) connected together. The main feature of an LC circuit is that it can resonate at a specific frequency, known as the resonant frequency, which is determined by the values of the inductor and the capacitor.

When energy is applied to the circuit, the capacitor stores electrical energy, and the inductor stores magnetic energy. These energies oscillate back and forth between the two components, creating a resonant frequency where the circuit can either filter out or amplify signals. In radios, LC circuits are commonly used to select or tune into a specific frequency, allowing operators to pick up only the desired signals while filtering out unwanted frequencies.

What occurs in an LC circuit at resonance? (G5A12)

A. Current and voltage are equal
B. Resistance is cancelled
C. The circuit radiates all its energy in the form of radio waves
D. Inductive reactance and capacitive reactance cancel

In an LC circuit at resonance, the inductive reactance (opposition to AC from the inductor) and the capacitive reactance (opposition to AC from the capacitor) cancel each other out. This happens because, at the resonant frequency, the reactance from the inductor and the capacitor are equal in magnitude but opposite in direction. As a result, their effects balance out, and the circuit allows the AC signal to pass through with minimal opposition.

LC circuits are used in tuning, filtering, and signal processing in radio because they allow precise control over which frequencies are received or transmitted.

Electrical Circuits

In electrical circuits, how components like capacitors, inductors, and resistors are arranged—either in series or parallel—affects the overall behavior and performance of the circuit. Understanding these arrangements is necessary as they impact signal processing, filtering, and tuning.

In a series circuit, components are connected one after another, so the same current flows through each component, but the voltage is divided among them.

In contrast, in a parallel circuit, components are connected across the same two points, so the voltage is the same across each component, but the current is divided.

Whether components are in series or parallel affects the total reactance, impedance, and overall circuit performance, influencing how your radio equipment handles different frequencies. This section will explore the key differences and how these configurations are used in practical radio circuits.

Series Electrical Circuit

Parallel Electrical Circuit

Combination Electrical Circuit

Series

Capacitance in Series

Capacitance refers to a capacitor's ability to store electrical energy in an electric field. Capacitors are widely used in radio circuits to filter signals, block DC currents, and tune circuits by controlling frequency response. When capacitors are connected in series or parallel, their total capacitance changes. Understanding how to calculate total capacitance in different configurations is needed to tune and adjust circuits.

When capacitors are connected in series, the total capacitance is always less than the smallest individual capacitor. The formula for calculating the total capacitance in series is the reciprocal method. This involves taking the reciprocal (1 divided by the value) of each capacitor's capacitance, adding those reciprocals together, and then taking the reciprocal of the total to find the overall capacitance. This approach ensures the proper calculation for capacitors working together in a series configuration.

Let's work through an exam question together.

What is the capacitance of three 100-microfarad capacitors connected in series? (G5C09)

A. 0.33 microfarads
B. 3.0 microfarads
C. 33.3 microfarads
D. 300 microfarads

First, we start with our formula:

$$\frac{1}{C_{total}} = \frac{1}{C_1} + \frac{1}{C_2} + \frac{1}{C_3}$$

Second, substitute values. For three 100-microfarad capacitors connected in series:

$$\frac{1}{C_{total}} = \frac{1}{100} + \frac{1}{100} + \frac{1}{100}$$

Third, solve the equation:

$$\frac{1}{C_{total}} = \frac{1}{100} + \frac{1}{100} + \frac{1}{100} = \frac{3}{100}$$

Now, take the reciprocal to find the total capacitance (flip the answer upside down):

$$\frac{1}{C_{total}} = \frac{100}{3} = 33.3 \ microfarads$$

So, the total capacitance of three 100-microfarad capacitors in series is 33.3 microfarads.

Keep in mind that when capacitors are connected in series, the total capacitance is lower than the smallest individual capacitor. In our example, the lowest capacitance was 100 microfarads, and our answer was 33.3.

Let's try another question.

What is the capacitance of a 20-microfarad capacitor connected in series with a 50-microfarad capacitor? (G5C12)

A. 0.07 microfarads
B. 14.3 microfarads
C. 70 microfarads
D. 1,000 microfarads

To calculate the total capacitance in a series configuration, you need to use the reciprocal formula:

$$\frac{1}{C_{total}} = \frac{1}{C_1} + \frac{1}{C_2}$$

For this question, you are given a 20-microfarad capacitor and a 50-microfarad capacitor. Using our formula, substitute in the values:

$$\frac{1}{C_{total}} = \frac{1}{20} + \frac{1}{50}$$

Solve by adding the fractions. Since they don't share a common denominator, it will be easier to add them as decimals.

$$\frac{1}{20} = 0.05 \; and \; \frac{1}{50} = 0.02$$

Which equals:

$$\frac{1}{C_{total}} = 0.05 + 0.02 = 0.07$$

Now, take the reciprocal to find the total capacitance:

$$\frac{1}{C_{total}} = \frac{1}{0.07} = 14.3 \; microfarads$$

So, the total capacitance of a 20-microfarad capacitor connected in series with a 50-microfarad capacitor is 14.3 microfarads. When capacitors are connected in series, the total capacitance (14.3 microfarads) is less than the smallest individual capacitor in the series (20 microfarads).

For the exam, remember that when capacitors are connected in series, their total capacitance is always less than the smallest capacitor in the series.

Inductance in Series

Inductance is the property of an inductor that allows it to store energy in a magnetic field when current flows through it. Inductors are components in radio circuits used for tuning and filtering signals. When designing or adjusting circuits, there are times when you may need to increase the total inductance. Depending on your goals, this can be achieved by modifying the configuration of inductors in the circuit, either in series or parallel.

Which of the following components should be added to an inductor to increase the inductance? (G5C14)

A. A capacitor in series
B. A capacitor in parallel
C. An inductor in parallel
D. An inductor in series

To increase the inductance in a circuit, you should <u>add another inductor in series</u> with the existing one. When inductors are connected in series, their inductances add together, increasing the total inductance. The formula is straightforward:

$$L_{total} = L_1 + L_2 + ...$$

For example, if you have a 10-millihenry inductor and add another 10-millihenry inductor in series, the total inductance becomes 20 millihenries.

To remember this, consider inductors in series as "stacking" their magnetic fields, increasing the total inductance. This is the opposite of capacitors in series, where the total capacitance decreases. Adding inductors in series is a simple and effective way to increase inductance in radio and electronic circuits.

Let's try a question.

> **What is the inductance of a circuit with a 20-millihenry inductor connected in series with a 50-millihenry inductor? (G5C11)**
>
> A. 7 millihenries
> B. 14.3 millihenries
> **C. 70 millihenries**
> D. 1,000 millihenries

When inductors are connected in series, the total inductance is simply the sum of the individual inductances. This happens because, in a series circuit, the inductors share the same current, and their magnetic fields combine, leading to a straightforward addition of their inductances.

Step-by-Step Calculation:

To calculate the total inductance for a circuit with a 20-millihenry inductor connected in series with a 50-millihenry inductor, you use the following formula:

$$L_{total} = L_1 + L_2$$

Substituting the given values:

$$L_{total} = 20 \text{ mH} + 50 \text{ mH}$$

Solve the formula:

$$L_{total} = 20 \text{ mH} + 50 \text{ mH} = 70 \text{ millihenries}$$

<u>The total inductance of a 20-millihenry inductor connected in series with a 50-millihenry inductor is 70 millihenries.</u>

Inductors connected in series are added together, making it a straightforward calculation to increase inductance in a circuit.

Resonance in Series

Resonance is a critical concept in electronics, especially in radio circuits. In an LC circuit, which consists of an inductor (L) and a capacitor (C), resonance occurs when the inductive reactance (the opposition to current flow by the inductor) equals the capacitive reactance (the opposition by the capacitor). When these reactances are equal, the circuit operates at its resonant frequency, a frequency at which the circuit behaves in a highly efficient manner.

What happens when inductive and capacitive reactance are equal in a series LC circuit? (G5A01)

A. Resonance causes impedance to be very high
B. Impedance is equal to the geometric mean of the inductance and capacitance
C. Resonance causes impedance to be very low
D. Impedance is equal to the arithmetic mean of the inductance and capacitance

The circuit reaches resonance when inductive and capacitive reactance are equal in a series LC circuit. At resonance, the opposing effects of inductors and capacitors cancel each other out, <u>causing the circuit's impedance (total opposition to current flow) to be very low.</u> This means the circuit allows maximum current to flow through it at the resonant frequency with very little resistance.

At this point, the LC circuit can easily pass resonant frequency signals while blocking signals of other frequencies. This property of low impedance at resonance makes LC circuits highly effective for tuning radios to desired frequencies or filtering out unwanted signals.

Consider resonance the "sweet spot" where the inductive and capacitive reactance cancel each other, resulting in low impedance and efficient energy transfer.

Parallel

Capacitance in Parallel

To increase the total capacitance in a circuit, you should <u>add another capacitor in parallel</u> with the existing one. When capacitors are connected in parallel, their capacitances add together. This is because each capacitor provides an additional path for storing electrical charge, effectively increasing the overall capacity of the circuit to store energy. The formula for total capacitance in parallel is:

$$C_{total} = C_1 + C_2 + ...$$

So, by adding another capacitor in parallel, you increase the total capacitance, allowing the circuit to store more energy.

Which of the following components should be added to a capacitor to increase the capacitance? (G5C13)

A. An inductor in series
B. An inductor in parallel
C. A capacitor in parallel
D. A capacitor in series

To remember this for the exam, think of parallel capacitors as "working together" to increase capacitance. This is different from capacitors in series, where the overall capacitance decreases. Adding capacitors in parallel can be helpful when fine-tuning filters or other components that rely on specific capacitance values.

Exam Time

> **What is the equivalent capacitance of two 5.0-nanofarad capacitors and one 750-picofarad capacitor connected in parallel? (G5C08)**
>
> A. 576.9 nanofarads
> B. 1,733 picofarads
> C. 3,583 picofarads
> **D. 10.750 nanofarads**

When capacitors are connected in parallel, their capacitances add together to form the total capacitance. In this case, you have two 5.0-nanofarad capacitors and one 750-picofarad capacitor (remember the two(2) 5.0 nanofarad capacitors. I didn't register the written word two the first time I did this! Also, note the different units: nanofarad and picofarad. We need to convert!

First, convert the 750 picofarads to nanofarads to calculate the total capacitance because the units need to match. Since 1 nanofarad equals 1,000 picofarads, 750 picofarads is equivalent to 0.750 nanofarads.

What is 750 picofarads in nanofarads?

$$750 \text{ pF} = 0.750 \text{ nF}$$

Moving to the Left the 'Big Number' Out Front Gets **SMALLER**

Now, you can add all the capacitances together:

$$C_{total} = 5.0 \text{ nF} + 5.0 \text{ nF} + 0.750 \text{ nF} = 10.750 \text{ nanofarads}$$

So, the equivalent capacitance of the three capacitors connected in parallel is 10.750 nanofarads.

Remember parallel capacitors as "adding up" their ability to store charge. This simple addition rule applies no matter how many capacitors you have in parallel, and it makes calculating the total capacitance straightforward.

Inductance in Parallel

Inductance is a fundamental property of electrical circuits, particularly those involving coils or inductors. It refers to the ability of an inductor to store energy in a magnetic field when current flows through it.

When inductors are connected in parallel, the total inductance is always less than the smallest individual inductor. The formula for calculating the total inductance in parallel is similar to how resistors are calculated in parallel:

$$\frac{1}{L_{total}} = \frac{1}{L_1} + \frac{1}{L_2} + \frac{1}{L_3}$$

What is the inductance of three 10-millihenry inductors connected in parallel? (G5C10)

A. 0.30 henries
B. 3.3 henries
C. 3.3 millihenries
D. 30 millihenries

By now, this math should be looking pretty straightforward!

For three 10-millihenry inductors connected in parallel:

$$\frac{1}{L_{total}} = \frac{1}{10} + \frac{1}{10} + \frac{1}{10} = \frac{3}{10}$$

Now, take the reciprocal to find the total inductance:

$$L_{total} = \frac{10}{3} = 3.3 \ millihenries$$

So, the total inductance of three 10-millihenry inductors in parallel is 3.3 millihenries.

To remember this for the exam, note that when inductors are connected in parallel, the total inductance is lower than the smallest inductor, and you use the reciprocal method to calculate the total value.

Resistors in Parallel

In a circuit with parallel resistors, the total current flowing through the circuit equals the sum of the currents flowing through each branch. This happens because, in a parallel configuration, each resistor provides an independent path for the current to flow. The more branches you have, the more paths there are for the current, so the total current increases as it splits among the different resistors.

The formula for total current (I_{total}) is:

$$I_{total} = I_1 + I_2 + ...$$

Each 'I' represents the current through an individual branch. Since the voltage across each resistor in parallel is the same, the current through each resistor depends on its resistance according to Ohm's Law.

How does the total current relate to the individual currents in a circuit of parallel resistors? (G5B02)

A. It equals the average of the branch currents
B. It decreases as more parallel branches are added to the circuit
C. It equals the sum of the currents through each branch
D. It is the sum of the reciprocal of each individual voltage drop

Think of parallel circuits as "splitting" the total current into separate branches. Each branch carries part of the total current, and when you add them together, you get the total current flowing into the circuit. This is different from series circuits, where the same current flows through all components.

Resistance in Parallel

When resistors are connected in parallel, the total resistance is always less than the smallest individual resistor. The formula for calculating total resistance (R_{total}) in a parallel circuit with two resistors is:

$$\frac{1}{R_{total}} = \frac{1}{R_1} + \frac{1}{R_2} + \frac{1}{R_3}$$

What is the approximate total resistance of a 100- and a 200-ohm resistor in parallel? (G5C04)

A. 300 ohms
B. 150 ohms
C. 75 ohms
D. 67 ohms

The two resistors have 100 ohms and 200 ohms values in this case.

Using the same type of math from the 'Capacitance in Series' section, we can solve the formula:

$$\frac{1}{R_{total}} = \frac{1}{100} + \frac{1}{200} = 0.01 + 0.005 = 0.015$$

Now, take the reciprocal to find the total resistance:

$$R_{total} = \frac{1}{0.015} \approx 66.7 \ ohms$$

So, the approximate total resistance of a 100-ohm and a 200-ohm resistor in parallel is 66.7 or 67 ohms.

To recall this for the exam, remember that for resistors in parallel, the total resistance is always lower than the smallest individual resistor, and the combined resistance can be calculated using the reciprocal formula.

Resistors & Resistance: How These Last Two Ideas Relate

Adding more resistors in parallel lowers the circuit's total resistance, allowing more total current to flow through it. As total resistance decreases, total current increases. Essentially, the more paths you give the current (by adding more resistors in parallel), the less resistance the circuit offers, and the more current will flow. So, these two concepts are linked: decreasing total resistance increases the total current in a parallel circuit.

Parallel Circuit

In a parallel circuit, the total current goes up when the total resistance goes down because the current has more paths to take through the circuit.

Let's answer another question.

What is the total resistance of a 10-, a 20-, and a 50-ohm resistor connected in parallel? (G5C03)

A. 5.9 ohms
B. 0.17 ohms
C. 17 ohms
D. 80 ohms

You use the reciprocal formula to find the total resistance of resistors connected in parallel. For three resistors with values of 10 ohms, 20 ohms, and 50 ohms, the formula is:

$$\frac{1}{R_{total}} = \frac{1}{R_1} + \frac{1}{R_2} + \frac{1}{R_3}$$

Substituting the resistor values:

$$\frac{1}{R_{total}} = \frac{1}{10} + \frac{1}{20} + \frac{1}{50}$$

Since the denominators aren't the same, let's switch to decimals.

$$\frac{1}{10} = 0.1 \ \ and \ \ \frac{1}{20} = 0.05 \ \ and \ \ \frac{1}{50} = 0.02$$

$$0.1 + 0.05 + 0.02 = 0.17$$

Now, take the reciprocal of 0.17 to find the total resistance:

$$R_{total} = \frac{1}{0.17} \approx 5.88 \ ohms$$

So, the total resistance of the 10-ohm, 20-ohm, and 50-ohm resistors connected in parallel is approximately 5.88, or rounded to 5.9 ohms, which is lower than the smallest resistor (10 ohms).

To remember this for the exam, note that the total resistance in a parallel circuit is always less than the smallest individual resistor. By using the reciprocal formula, you can calculate how the total resistance combines in parallel circuits.

Transformers

A transformer is an electrical device used to transfer electrical energy between two or more circuits through electromagnetic induction. It consists of two or more coils of wire, known as windings, wrapped around a shared core, typically made of iron. When alternating current flows through one coil (the primary winding), it creates a magnetic field that induces a voltage in the other coil (the secondary winding). This allows the transformer to increase or decrease voltage levels while maintaining the same frequency.

Transformers allow impedance matching, voltage regulation, and isolation between different circuit parts. For example, transformers help match the impedance between a transmitter and an antenna, ensuring that power is transferred efficiently without signal loss or reflections. By adjusting the voltage and current ratios, transformers play a role in optimizing the performance of radio equipment.

When an AC voltage source is connected to the primary winding of a transformer, it creates a changing magnetic field around the winding. This changing magnetic field induces a voltage in the secondary winding through mutual inductance. Mutual inductance occurs because the magnetic field generated by the current in the primary winding interacts with the secondary winding, causing a voltage to appear across it. This is the fundamental principle of how transformers work, allowing them to transfer electrical energy from one circuit to another without a direct electrical connection.

What causes a voltage to appear across the secondary winding of a transformer when an AC voltage source is connected across its primary winding? (G5C01)

A. Capacitive coupling
B. Displacement current coupling
C. Mutual inductance
D. Mutual capacitance

Mutual inductance is the "magnetic link" between the primary and secondary windings. The AC in the primary winding generates a magnetic field, which "induces" or transfers voltage into the secondary winding through this magnetic connection. This is why transformers can step up or down voltages in AC circuits, making them key in radio and electrical systems.

Transformer Design

Step-Up Transformer

In a voltage step-up transformer, the primary winding wire is usually larger than the secondary winding wire to handle the higher current flowing through it. When a transformer steps up the voltage, it simultaneously steps down the current. This means the primary winding, which operates at a lower voltage, must carry a higher current, while the secondary winding, operating at a higher voltage, carries a lower current. Since larger wire sizes are better suited to handle higher currents without overheating, the primary winding is made with thicker wire to manage the increased current safely.

Why is the primary winding wire of a voltage step-up transformer usually a larger size than that of the secondary winding? (G5C05)

A. To improve the coupling between the primary and secondary
B. To accommodate the higher current of the primary
C. To prevent parasitic oscillations due to resistive losses in the primary
D. To ensure that the volume of the primary winding is equal to the volume of the secondary winding

In a step-up transformer, the voltage goes up on the secondary side, but the current goes down. Therefore, the primary winding needs to handle more current, requiring a thicker wire, while the secondary winding can use a thinner wire since it carries less current. This balance ensures the transformer operates efficiently and safely.

Step-Down Transformer

A step-down transformer reduces voltage from a higher level to a lower level. It does this by having fewer turns (windings) on the secondary coil than on the primary coil. When alternating current is applied to the primary winding, the transformer decreases the voltage proportionally according to the turns ratio.

What is the output voltage if an input signal is applied to the secondary winding of a 4:1 voltage step-down transformer instead of the primary winding? (G5C02)

A. The input voltage is multiplied by 4
B. The input voltage is divided by 4
C. Additional resistance must be added in series with the primary to prevent overload
D. Additional resistance must be added in parallel with the secondary to prevent overload

Ok, so this is a tricky question because they are asking it "in reverse."

In a voltage step-down transformer with a 4:1 turns ratio, the transformer is designed to reduce the voltage when applied to the primary winding. However, suppose you reverse the setup and apply the input signal to the secondary winding instead of the primary. In that case, the transformer acts in reverse, becoming a step-up transformer. As a result, the input voltage is multiplied by the same ratio, which in this case is 4.

For example, apply 10 volts to the secondary winding of this 4:1 step-down transformer. The output voltage from the primary winding will be 40 volts. This happens because the transformer boosts the voltage by the same factor (4) it would normally step down when used in the conventional configuration. To recall this for the exam, remember that switching the input to the secondary winding reverses the transformer's function, multiplying the voltage by the turns ratio.

Transformer Math

When understanding the math associated with transformers, breaking things down step by step is important, especially for beginners. Transformers work based on the principle of proportionality between voltage, current, and the number of windings in the coils.

The formulas revolve around these relationships.

Voltage and Turns Ratio: The voltage in a transformer is directly proportional to the number of turns (or windings) in the coils. This means:

$$\frac{V_{primary}}{V_{secondary}} = \frac{N_{primary}}{N_{secondary}}$$

Where Vprimary is the voltage in the primary winding, Vsecondary is the voltage in the secondary winding, and Nprimary and Nsecondary are the number of turns in the primary and secondary windings, respectively. If the secondary coil has more turns than the primary, the voltage increases (step-up transformer). If it has fewer turns, the voltage decreases (step-down transformer).

Current and Turns Ratio: The current behaves inversely to the number of turns in the transformer. When the voltage is stepped up, the current is stepped down, and vice versa. The formula for current is:

$$\frac{I_{primary}}{I_{secondary}} = \frac{N_{secondary}}{N_{primary}}$$

Where Iprimary is the current in the primary winding, and Isecondary is the current in the secondary winding. So, the current decreases as the voltage increases in the secondary coil.

The takeaway for beginners is that voltage and current are related to the number of turns in the windings, and they trade off. If one goes up, the other goes down, and the math helps us calculate exactly how much! In the upcoming questions, I'll break down each concept so that even those new to this topic can easily follow along.

Let's try some questions.

What is the voltage output of a transformer with a 500-turn primary and a 1500-turn secondary when 120 VAC is applied to the primary? (G5C06)

A. 360 volts
B. 120 volts
C. 40 volts
D. 25.5 volts

The information from the previous section provides the basic understanding needed to answer this question. It uses the relationship between the number of turns in the transformer windings and the voltage, known as the voltage turns ratio.

Here's how to solve this:

$$\frac{V_{primary}}{V_{secondary}} = \frac{N_{primary}}{N_{secondary}}$$

This question is asking about voltage, so we'll use the transformer turns ratio for voltage.

Where:

- $V_{primary}$ is the voltage applied to the primary winding

- $V_{secondary}$ is the voltage output of the secondary winding

- $N_{primary}$ is the number of turns in the primary winding

- $N_{secondary}$ is the number of turns in the secondary winding

Step-by-step solution:

First, identify the values given:

- Primary turns ($N_{primary}$) = 500 turns

- Secondary turns ($N_{secondary}$) = 1500 turns

- Primary voltage ($V_{primary}$) = 120 VAC (volts alternating current)

Second, set up the formula:

$$\frac{120}{V_{secondary}} = \frac{500}{1{,}500}$$

Finally, cross multiply to eliminate the denominator: 120 * 1500 = 180,000, then divide by 500 = 360

Answer: The voltage output of the secondary winding is 360 VAC.

Since the secondary winding has three times more turns than the primary winding, the voltage in the secondary will be three times higher than the primary voltage. This transformer is a step-up transformer, increasing the voltage from 120 VAC to 360 VAC.

Transformers & Impedance

In radio systems, impedance matching is necessary to ensure efficient signal transfer between components, such as between an antenna and a transmission line. When the impedance of these components doesn't match, a portion of the signal is reflected back, causing power loss and reduced performance. This is where transformers come into play. By adjusting the ratio of turns in the primary and secondary windings, a transformer can match the impedance of two components, allowing maximum power transfer with minimal loss.

Transformers work by adjusting the voltage and current ratios, and since impedance is related to both, transformers can also change the impedance between circuits. For example, suppose an antenna has a 600-ohm impedance and needs to connect to a 50-ohm coaxial cable. In that case, a transformer can be used to "step down" the impedance. This is done by choosing a transformer with the correct turns ratio—in this case, about 3.46:1—to match the higher antenna impedance to the lower cable impedance, ensuring efficient signal transfer and system performance.

What transformer turns ratio matches an antenna's 600-ohm feed point impedance to a 50-ohm coaxial cable? (G5C07)

A. 3.5 to 1
B. 12 to 1
C. 24 to 1
D. 144 to 1

So, what's the math behind this, and how do we get 3.46 or rounded up to 3.5?

You need a transformer that adjusts the impedance between the two to match an antenna's 600-ohm feed point impedance to a 50-ohm coaxial cable. The turns ratio of a transformer that matches impedances is determined by the square root of the impedance ratio between the feed point and the coaxial cable. The formula is:

$$Turns\ Ratio = \sqrt{\frac{Z_{primary}}{Z_{secondary}}}$$

Where:

- Zprimary is the 600-ohm impedance of the antenna feed point.

- Zsecondary is the 50-ohm impedance of the coaxial cable.

Using the formula:

$$Turns\ Ratio = \sqrt{\frac{Z_{primary}}{Z_{secondary}}} = \sqrt{\frac{600}{50}} = \sqrt{12} \approx 3.46\ or\ 3.5\ to\ 1$$

This means the transformer needs a turns ratio of about 3.46, rounded to 3.5:1, to effectively match the 600-ohm feed point to the 50-ohm coaxial cable. This allows maximum power transfer between the antenna and the transmission line with minimal signal loss due to impedance mismatch.

Chapter Summary

In this chapter on Electrical Principles, we've explored some advanced concepts and math that form the backbone of radio and electronics. Beginning with electrical components and moving through configurations like series and parallel circuits, you have gained insight into how these elements combine to build functional circuits.

We then examined how reactance affects current flow in AC circuits, leading to a deeper understanding of impedance and admittance. These principles are crucial for managing circuit behavior, optimizing energy flow, and achieving resonance—a phenomenon at the core of frequency selection and tuning in radio applications. Finally, transformers illustrated a practical application of these principles, enabling energy transfer and signal modification across circuits. Together, these topics build a comprehensive foundation for understanding the electrical mechanics of radio communication.

Chapter 8

Power

POWER IS A FUNDAMENTAL concept in electronics and radio communication, directly influencing how efficiently your equipment operates and how far your signals can travel. In simple terms, power is the rate at which energy is used or transmitted. It's measured in watts (W) and calculated by multiplying voltage and current. Understanding power is critical for radio operators because it affects signal strength, transmission range, and compliance with legal transmission limits.

Power is especially important in radio systems because it determines how well you can send and receive signals. Whether tuning your transmitter or adjusting your antenna, knowing how to manage and calculate power ensures that your signal is strong enough to communicate clearly without exceeding legal limits or causing interference. This section will dive into the various types of power and explain how to calculate and manage power.

Power Formula

When we talk about consuming watts of electrical power, we're referring to the amount of energy a device or circuit uses over time. A watt is a unit of power that measures the rate at which energy is consumed or produced. In electrical terms, power (in watts) is calculated by multiplying the voltage (in volts) by the current (in amperes). The formula for power is:

Power	Current	Voltage
$\widehat{P}$ / I | E	P / $\widehat{I}$ | E	P / I | $\widehat{E}$
P = E × I	I = P / E	E = P / I

$$P = I \times E$$

In a radio transmitter, the watts consumed determine how much energy is drawn from the power source to generate radio signals. The more watts a transmitter consumes, the more powerful the signal it can send, which often translates to greater range and clarity.

However, managing power consumption is also important, as too much power can lead to overheating, wear on equipment, or exceeding legal power limits. By understanding how watts relate to your equipment's performance, you can better balance power needs with efficiency and longevity in your radio setup.

Let's try some exam questions.

Here is a basic one to start with.

How many watts of electrical power are consumed by a 12 VDC light bulb that draws 0.2 amperes? (G5B04)

A. 2.4 watts
B. 24 watts
C. 6 watts
D. 60 watts

Remembering back to our Technician's test, to calculate the number of watts consumed by a 12 VDC light bulb that draws 0.2 amperes, you can use the basic power formula because we want Power (P) and we have Voltage (V) and Current (I)

$$P = I \times E$$

Where:

- P is the power in watts,

- E is the voltage (12 VDC),

- I is the current in amperes (0.2 A).

Substitute the given values into the formula:

$$P = 12 \times 0.2 = 2.4 \, watts$$

So, the light bulb consumes 2.4 watts of electrical power.

Next, let's try a couple more advanced problems.

How many watts of electrical power are consumed if 400 VDC is supplied to an 800-ohm load? (G5B03)

A. 0.5 watts
B. 200 watts
C. 400 watts
D. 3200 watts

First, VDC.

VDC stands for Volts Direct Current, which refers to the voltage supplied in a direct current (DC) circuit. In DC circuits, the current flows in only one direction, as opposed to alternating current (AC), where the direction of the current changes periodically.

VDC is used to specify the voltage level in DC systems. For example, a 12 VDC power supply provides a steady 12 volts in a single direction, which is typical for devices like batteries or power supplies for electronics.

Understanding VDC is important because it helps differentiate between direct current (constant, one-way flow of electrons) and alternating current (where the flow of electrons changes direction). In many radio systems, DC voltage powers the equipment, making VDC a critical concept.

But honestly, VDC vs DC still wasn't clear to me.

The terms **VDC** (Volts Direct Current) and **DC** (Direct Current) are closely related but have slightly different uses. **DC** refers to the type of current where electricity flows in a constant direction, while **VDC** specifically measures the voltage in a DC circuit. For example, a power source may supply DC power, but its voltage would be described as 12 VDC or 13.8 VDC. This distinction is important for precision, as many devices, including ham radios, require specific voltage levels for safe and efficient operation. Using "VDC" clarifies that the voltage is from a direct current source, avoiding confusion with AC voltage (VAC).

So now, to calculate how many watts of electrical power are consumed when 400 VDC is supplied to an 800-ohm load, you can use Ohm's Law and the power formula.

Voltage	Current	Resistance
$E = I . R$	$I = E/R$	$R = E / I$

Step 1: There are two key formulas you need:

Ohm's Law

Ohm's Law: $E = I \times R$

- E: Voltage in volts

- I: Current in amperes

- R: Resistance in ohms

Rearranging this formula can help us find current (I) if voltage (V) and resistance (R) are known:

$$I = \frac{E}{R}$$

Power Formula

Power Formula: $P = E \times I$

- P: Power in watts

- E: Voltage in volts

- I: Current in amperes

Step 2: Combine the Formulas

From **Ohm's Law**, we know that:

$$I = \frac{E}{R}$$

substitue this expression for I into the Power Formula

$$P = E \times \frac{E}{R}$$

Simplify:

$$P = \frac{E^2}{R}$$

This combined formula allows us to calculate power directly when we know the voltage (V) and resistance (R).

You are now left with two options: One, memorize another formula, or two, just remember Power and Ohms and how to combine them.

Step 3: Solve the Problem

$$P = \frac{E^2}{R}$$

Where:

- P is the power in watts,

- E is the voltage (400 VDC),

- R is the resistance (800 ohms).

Now, substitute the given values into the formula:

$$P = \frac{400^2}{800} = \frac{160,000}{800} = 200 \; watts$$

So, 200 watts of electrical power are consumed by the 800-ohm load when 400 VDC is supplied.

How many watts are consumed when a current of 7.0 milliamperes flows through a 1,250-ohm resistance? (G5B05)

A. Approximately 61 milliwatts
B. Approximately 61 watts
C. Approximately 11 milliwatts
D. Approximately 11 watts

Again, let's start with our two known formulas. **Ohm's Law** and the **formula for power**.

Step 1: These are two key formulas you need:

Ohm's Law:

Ohm's Law: $E = I \times R$

- E: Voltage in volts

- I: Current in amperes

- R: Resistance in ohms

Power Formula:

Power Formula: $P = E \times I$

- P: Power in watts

- E: Voltage in volts

- I: Current in amperes

Step 2: Combine the Formulas

From **Ohm's Law**, we know that:

$$E = I \times R$$

substitue this expression for E into the Power Formula

$$P = (I \times R) \times I$$

Simplify:

$$P = I^2 \times R$$

Why did we do it this way? Because we are being asked for Power and have been given current and resistance. So, we need a formula that only uses current and resistance.

This combined formula allows us to calculate power directly when we know the voltage (E) and resistance (R).

You are now left with two options. One, memorize another formula, or two, just remember Power and Ohms and how to combine them together.

Step 3: Solve the Problem

$$P = I^2 \times R$$

Where:

- P is the power in watts (?)

- I is the current (7.0 milliamps)—this is given as milliamps, but we want to use amps (the base unit) in our math. So 7.0 milliamps = 0.007 amps or 7.0×10^{-3} amps.

- R is the resistance (1,250 ohms)

Now, substitute the given values into the formula:

$$P = (7.0 \times 10^{-3})^2 \times 1,250$$

Do the math:

$$(7.0 \times 10^{-3})^2 = 49 \times 10^{-6} = 0.000049$$

$$P = 0.000049 \times 1,250 = 0.06125 \ watts$$

Final Answer

The power consumed is 0.0612 watts, or converted to a whole number, 61.25 milliwatts are consumed when a current of 7.0 milliamps flows through a 1,250 ohm resistance.

Batteries

Batteries with Low Internal Resistance

One of the advantages of batteries with low internal resistance is their ability to deliver high discharge current. Internal resistance is the inherent opposition within a battery to the flow of current. When a battery has low internal resistance, it can provide more current to power devices efficiently without losing too much energy as heat. This makes such batteries ideal for applications that require high power output, such as in portable ham radio setups, where reliable and strong energy flow is needed to keep communications running smoothly.

What is an advantage of batteries with low internal resistance? (G6A02)

A. Long life
B. High discharge current
C. High voltage
D. Rapid recharge

Batteries with low internal resistance ensure that the radio station operates smoothly, especially during high-demand activities like transmitting at higher power levels. When the internal resistance is low, the battery can maintain stable voltage levels under load, resulting in better performance and longer operating times for radio equipment.

Batteries with low internal resistance can provide higher discharge currents, making them better suited for high-power applications and ensuring consistent and reliable radio performance. Understanding this concept lets you choose the correct battery for your ham radio equipment.

Lead-Acid Battery Discharge and Lifespan

To maximize the lifespan of a standard 12-volt lead-acid battery, it's important to avoid discharging it too deeply. The minimum allowable discharge voltage for these batteries is 10.5 volts. This means that when the battery's voltage drops to 10.5 volts under load, it's time to stop using it and recharge it. Discharging below this point can cause permanent damage, reducing the battery's lifespan and ability to hold a charge in the future.

What is the minimum allowable discharge voltage for maximum life of a standard 12-volt lead-acid battery? (G6A01)

A. 6 volts
B. 8.5 volts
C. 10.5 volts
D. 12 volts

Understanding how to manage the discharge of a lead-acid battery is key for radio operators, especially during portable or emergency operations. Keeping the battery above 10.5 volts ensures optimal performance. It avoids damaging the battery, which would affect your ability to operate your radio reliably over time.

Decibels and Gain

Decibels (dB) are a fundamental unit of measurement used to quantify signal strength and gain. A decibel is a logarithmic unit that measures the ratio between two values, such as power levels or signal strengths. This logarithmic scale is handy because it can represent a wide range of values in a more manageable way. For instance, rather than dealing with large numbers, decibels allow you to express these quantities in simpler terms. In ham radio, decibels are used to compare signal levels, assess antenna gain, and evaluate amplifier performance. Understanding dB helps you make informed decisions about your equipment and operating conditions, ensuring optimal performance.

Gain is another critical concept in radio communication. It refers to the increase in signal strength achieved by an amplifier or antenna. Gain can be categorized into voltage, current, and power gain. Voltage gain refers to the increase in voltage, current gain to the rise in current, and power gain to the increase in power output. High gain is vital for improving signal reach and clarity, especially in weak signal conditions. It allows your transmissions to travel farther and be received more clearly, which is particularly important for DXing and other long-distance communications.

What dB change represents a factor of two increase or decrease in power? (G5B01)

A. Approximately 2 dB
B. Approximately 3 dB
C. Approximately 6 dB
D. Approximately 9 dB

A change of approximately 3 dB (decibels) represents either a doubling or halving of power – a factor of two increase). Specifically, a +3 dB change means the power has increased by a factor of two, while a -3 dB change means the power has decreased by half. This relationship is important because decibels are a logarithmic scale used to express ratios of power or voltage in a way that is easier to manage, especially when dealing with large ranges of values.

The 3 dB change in power comes from the logarithmic relationship between power ratios and decibels. The formula to calculate decibels when comparing two power levels is:

$$dB = 10 \times log_{10}\left(\frac{P_2}{P_1}\right)$$

Hopefully, this looks familiar from your Technician's exam!

Where:

- P2 is the final power level

- P1 is the initial power level

To understand why 3 dB represents a doubling or halving of power, let's consider the cases where the power ratio is 2 (doubling) or 0.5 (halving):

Doubling Power: $dB = 10 \times log_{10}(2) \approx 10 \times 0.3010 = 3\ dB$

Halving Power: $dB = 10 \times log_{10}(0.5) \approx 10 \times (-0.3010) = -3\ dB$

So, when the power increases by a factor of 2, the result is approximately +3 dB, and when the power decreases by half, the result is approximately -3 dB. This logarithmic relationship allows us to easily express large changes in power with relatively small dB values.

Another math question.

What percentage of power loss is equivalent to a loss of 1 dB? (G5B10)

A. 10.9 percent
B. 12.2 percent
C. 20.6 percent
D. 25.9 percent

A loss of 1 dB corresponds to a power loss of approximately 20.6 percent. This means that if you decrease the power of a signal by 1 dB, about 20.6% of the original power is lost, leaving you with roughly 79.4% of the original power. The decibel scale is logarithmic, which means that even small changes in dB represent significant changes in power.

To calculate the percentage of power loss corresponding to a 1 dB change, we can use the formula that relates decibels (dB) to the ratio of power levels:

$$dB = 10 \times log_{10}\left(\frac{P_2}{P_1}\right)$$

Where:

- P2 is the final power level.

- P1 is the initial power level.

Step-by-Step Calculation:

First, Set dB to -1 for a 1 dB loss:

$$-1 = 10 \times log_{10}\left(\frac{P_2}{P_1}\right)$$

Next, Divide by 10 to isolate the logarithm:

$$-0.1 = log_{10}\left(\frac{P_2}{P_1}\right)$$

Then, remove the logarithm by taking the inverse (antilog) of both sides:

$$\left(\frac{P_2}{P_1}\right) = 10^{-0.1} \approx 0.794$$

This means the final power (P2) is 79.4% of the initial power (P1) after a 1 dB loss.

Finally, calculate the percentage of power loss:

$$Power \; Loss \; = 100\% - 79.4\% = 20.6\%$$

So, a 1 dB loss corresponds to a 20.6% decrease in power.

Practical applications of dB and gain are numerous. Measuring antenna gain is one everyday use. Antenna gain is often expressed in dB, comparing the antenna's performance to a standard reference, such as an isotropic radiator (dBi) or a dipole antenna (dBd). For instance, an antenna with a gain of 6 dBd indicates it performs better than a standard dipole by 6 dB. Assessing amplifier performance is another application. By calculating the increase in signal strength provided by an amplifier, you can determine its effectiveness and suitability for your needs. Signal reports also frequently use dB to quantify improvements or degradations in signal strength. For example, an operator might report that your signal improved by 3 dB after you adjusted your antenna, indicating a noticeable enhancement in signal clarity.

Decibels and gain are concepts that allow you to measure and improve signal strength and quality, ensuring clear and effective transmissions. Whether comparing antennas, assessing amplifiers, or interpreting signal reports, a solid grasp of dB and gain empowers you to make informed decisions and enhance your overall operating experience.

Link Budget

A link budget is an essential calculation used to predict the performance of a communication link, especially over long distances. It accounts for all the factors that affect the signal from the transmitter to the receiver. By considering transmit power, antenna gains, and various system losses (such as cable losses or environmental factors), a link budget helps operators determine if the signal will be strong enough for reliable communication. Understanding link budgets is necessary to designing effective communication systems and ensuring signals are received clearly over the desired range.

A link budget is the calculation of the total power available for communication at the receiver's end. It starts with the transmit power, adds any gains provided by the transmitting and receiving antennas, and then subtracts the losses caused by factors like cable losses, free-space path loss, or atmospheric conditions. The result gives an estimate of how much signal power will be left when it reaches the receiver.

> **What is a link budget? (G8A13)**
>
> A. The financial costs associated with operating a radio link
> B. The sum of antenna gains minus system losses
> **C. The sum of transmit power and antenna gains minus system losses as seen at the receiver**
> D. The difference between transmit power and receiver sensitivity

Measuring a link budget in real life involves testing and analyzing each component of the communication link. Start by measuring transmitter power with a power meter and accounting for cable losses. Verify antenna gain using manufacturer specs or field tests. Calculate or measure path loss using the free space formula or real-world tools like signal strength meters. Assess other losses, such as coaxial cable or environmental factors, using a network analyzer and online attenuation calculators. Finally, measure the received signal strength with an S-meter or spectrum analyzer and compare it to the receiver's sensitivity to ensure reliable communication. By combining theoretical calculations with real-world testing, you can optimize your station for consistent performance.

To remember this for the exam, think of a link budget as a balance sheet for your signal. You start with the signal's initial strength, account for any improvements (gains), and subtract the losses the signal encounters during transmission. This calculation helps ensure that your system has enough power to communicate effectively, especially over long distances.

Link Margin

Link margin helps ensure a reliable connection between a transmitter and receiver. It represents the "buffer" or extra power the received signal has above the minimum required level for proper reception. Environmental factors, interference, and equipment aging can reduce signal strength, and having a sufficient margin helps maintain the quality of the communication link even when conditions degrade. A higher link margin means more reliable communication, especially over long distances or in noisy environments.

Link margin is the difference between the received power level and the minimum required signal level needed for the receiver to function properly. In other words, the extra signal strength ensures reliable communication. Suppose the received power is only slightly above the required level. In that case, the link margin is small, and interference or other factors could easily disrupt the connection. A larger link margin provides more room for error, helping to maintain a stable connection even in challenging conditions.

> **What is link margin? (G8A14)**
>
> A. The opposite of fade margin
> **B. The difference between received power level and minimum required signal level at the input to the receiver**
> C. Transmit power minus receiver sensitivity
> D. Receiver sensitivity plus 3 dB

To remember this for the exam, think of the link margin as a safety net—it's the extra power you have beyond what's minimally required to keep the communication link working reliably. This concept is key to designing systems that must function well even when faced with signal degradation or interference.

Chapter Summary

Understanding power principles is vital for managing and optimizing radio communication systems. Each concept ensures efficient signal transmission, from choosing batteries as power sources to the detailed calculations of decibels and gain. Radio operators can account for and compensate for losses by applying link budget and link margin principles, achieving clearer and more reliable connections. Mastery of these power fundamentals provides a solid foundation for advancing in the world of radio, where precise power management is critical for effective communication across distances and conditions.

The Physical Radio Station

Your radio station is the hub of all amateur radio activities, where theory meets practice. Whether you're assembling your first setup or optimizing an existing one, understanding the physical components of your station is key to smooth and reliable operation. Each element, from antennas and feedlines to power sources and circuit design, plays a critical role in ensuring clear communication and efficient performance.

In this section, we'll discuss the technical standards, equipment essentials, and best practices for building, maintaining, and upgrading your radio station. This will empower you to create a setup that meets your needs and keeps you on the air with confidence, all while fully preparing for your General exam.

Chapter 9

Equipment

Upgrading to a General Class license unlocks exciting new privileges but demands a deeper understanding of the equipment that powers your station. This chapter dives into the critical components of a functional ham radio station to fully prepare for the exam and the challenges of expanded operation.

At the heart of your station is the transceiver, your primary tool for transmitting and receiving signals. Choosing a transceiver that covers the General Class frequency bands and offers features like Digital Signal Processing (DSP) ensures optimal performance and clarity. Equally important is a reliable antenna and feedline setup—your connection to the world. From dipoles to Yagi antennas, understanding these components and their installation is key to maximizing signal strength and communication success. Let's explore these topics to help you operate confidently and effectively while preparing for your exam.

Transceiver

Transceivers are the heart of any ham radio station, integrating transmitting and receiving functions into one device. Whether operating from a permanent base station or on the go, selecting the right transceiver is critical for effective communication. Base station transceivers offer high power and advanced features but are larger and pricier. Mobile and handheld transceivers provide flexibility and portability but may lack the robust capabilities of base units.

Three standout models combine performance and reliability for General-class operators. The Icom IC-7300 is highly regarded for its advanced features, including a high-resolution touchscreen, built-in DSP, and an automatic antenna tuner covering all HF bands. The Yaesu FT-991A offers versatility, covering HF, VHF, and UHF bands, with a compact design and easy-to-use interface suited for both base and mobile operations. The Kenwood TS-590SG delivers reliability and user-friendly operation for exceptional receiver performance and advanced filtering, making it a strong choice for serious operators.

By understanding the strengths of these transceiver options, you can confidently choose the best fit for your General Class privileges and communication goals.

Checklist: Selecting Your Transceiver

- **Type**: Decide between base station, mobile, and handheld transceivers based on your operating preferences and space.

- **Frequency Coverage**: Ensure the transceiver covers all HF bands available to General Class operators.

- **Power Output:** Look for adjustable power settings to match different operating conditions.

- **User Interface**: Choose a model with a clear display, intuitive menus, and easily accessible controls.

- **Additional Features**: Consider DSP, automatic antenna tuners, and memory channels for enhanced performance.

Regular calibration and maintenance are required to optimize performance. Proper alignment prevents interference and ensures reliable operation, while external filters and preamplifiers enhance signal quality. High-pass and low-pass filters block unwanted frequencies, and preamplifiers boost weak signals. Additionally, keeping connectors clean, inspecting cables, and applying firmware updates can improve performance and extend the life of your equipment.

Voice Operated Exchange

In amateur radio, there are two common ways to control a radio's transmission: Voice Operated Exchange (VOX) and Push-To-Talk (PTT). These methods allow operators to switch between listening and transmitting. PTT requires the operator to manually press a button when they wish to transmit. VOX automatically triggers the transmitter when it detects the operator's voice. Each method has advantages and depends on the situation and the operator's preferences.

The key difference between VOX operation and PTT operation is that VOX allows "hands-free" operation. With VOX enabled, the radio begins transmitting as soon as it detects the operator's voice without needing to press a button. This can be especially useful when the operator needs to use both hands for something else, like when using a computer, logging contacts, or adjusting equipment.

Which of the following statements is true of VOX operation versus PTT operation? (G2A10)

A. The received signal is more natural sounding
B. It allows "hands-free" operation
C. It occupies less bandwidth
D. It provides more power output

In contrast, PTT operation requires the operator to press and hold a button every time they want to transmit, which provides more manual control but requires active engagement. The convenience of VOX can be balanced by its sensitivity to background noise, which may accidentally trigger transmissions, so operators should adjust VOX settings accordingly.

Two-Tone Test

A two-tone test is a standard method for evaluating the linearity and performance of a transmitter's modulation and amplifier stages. This test injects two pure, non-harmonic audio signals/ tones of equal amplitude into the transmitter's input, typically through a microphone or audio interface. These tones are chosen so their frequencies fall within the transmitter's bandwidth but do not interfere with each other.

When the transmitter amplifies the signal, its output is analyzed using a spectrum analyzer or oscilloscope. In an ideal scenario, the output will show only the original tones and their sum and difference frequencies (the sidebands) without additional distortion or spurious signals. However, if the transmitter exhibits non-linearity or overdrive, unwanted harmonics or intermodulation products will appear, indicating distortion.

This test is critical for ensuring the transmitter delivers a clean signal, minimizing interference with other users on the band. It helps operators identify and correct issues like improper gain settings or amplifier inefficiencies, ensuring compliance with FCC standards and optimal on-air performance.

What signals are used to conduct a two-tone test? (G4B07)

A. Two audio signals of the same frequency shifted 90 degrees
B. Two non-harmonically related audio signals
C. Two swept frequency tones
D. Two audio frequency range square wave signals of equal amplitude

The critical transmitter performance parameter analyzed by the two-tone test is linearity. Linearity ensures that the transmitted signals remain clean and without distortion, especially in amplitude modulation and single-sideband (SSB) operations. If the transmitter is non-linear, the output will include intermodulation products—unwanted signals generated by the interaction of the two tones, which can cause interference and degrade signal quality.

What transmitter performance parameter does a two-tone test analyze? (G4B08)

A. Linearity
B. Percentage of suppression of the carrier and undesired sideband for SSB
C. Percentage of frequency modulation
D. Percentage of carrier phase shift

To perform a two-tone test, combine two pure audio tones (e.g., 700 Hz and 1900 Hz) of equal amplitude using a signal generator and inject them into the transmitter's audio input. Connect the transmitter to a dummy load and begin transmitting. A spectrum analyzer or oscilloscope is used to observe the output signal at the dummy load, checking for clean tones and sidebands without intermodulation distortion or harmonics. Adjust the transmitter's audio gain or drive levels and retest if unwanted signals are present. This test ensures the transmitter operates linearly, producing a clean signal while meeting FCC standards. Always use a dummy load and proper equipment for safe and accurate results.

The two-tone test helps amateur operators analyze the linearity of their transmitters by introducing two different audio tones. The output is checked to ensure that no distortion or unwanted frequencies are produced, as non-linear operation can affect communication quality. The goal is to ensure clean, reliable transmissions.

Power Amplifiers

Automatic Level Control

Automatic Level Control (ALC) is a feature used with RF power amplifiers to prevent excessive drive, which could damage the amplifier or cause signal distortion. The ALC system works by monitoring the amplifier's input power and, if the drive signal becomes too strong, reducing it to maintain safe operating levels. This ensures that the amplifier functions efficiently and avoids pushing the equipment beyond its designed limits. Without ALC, transmitting with excessive drive could lead to overheating, distortion of the transmitted signal, and interference with other users on the frequency.

Why is automatic level control (ALC) used with an RF power amplifier? (G4A05)

A. To balance the transmitter audio frequency response
B. To reduce harmonic radiation
C. To prevent excessive drive
D. To increase overall efficiency

Side note: Drive refers to the input signal provided to an RF power amplifier, typically originating from a transceiver or exciter. The drive level determines how much power the amplifier amplifies and outputs.

However, when transmitting AFSK (Audio Frequency Shift Keying) data signals, the ALC system should be disabled because its action can distort the signal. AFSK signals rely on precise frequency shifts to encode data. If the ALC kicks in to reduce drive, it can alter the intended signal, resulting in data errors or poor transmission quality. Avoiding ALC when handling such digital modes is important to ensure signal clarity and reliable data transmission.

Why should the ALC system be inactive when transmitting AFSK data signals? (G4A11)

A. ALC will invert the modulation of the AFSK mode
B. The ALC action distorts the signal
C. When using digital modes, too much ALC activity can cause the transmitter to overheat
D. All these choices are correct

While ALC helps keep RF power amplifiers safe from overdrive, it must be used carefully, especially in digital data modes like AFSK, where precision is needed.

Vacuum-Tube RF Power Amplifier

When operating a vacuum-tube RF power amplifier, understanding how to properly adjust the TUNE and LOAD controls is crucial for optimal performance and avoiding damage. The TUNE control helps match the amplifier's internal circuitry to the antenna system, optimizing power transfer. A pronounced dip in plate current identifies the correct setting of this control. This dip indicates that the amplifier is tuned correctly, allowing maximum power output with minimal stress on the components.

What is the effect on plate current of the correct setting of a vacuum-tube RF power amplifier's TUNE control? (G4A04)

A. A pronounced peak
B. A pronounced dip
C. No change will be observed
D. A slow, rhythmic oscillation

Similarly, the LOAD or COUPLING control is used to adjust the amplifier's power output without exceeding the maximum allowable plate current. This setting ensures that the amplifier operates efficiently while staying within safe limits. By carefully adjusting this control, you achieve the desired power output without exceeding the maximum allowable plate current.

What is the correct adjustment for the LOAD or COUPLING control of a vacuum tube RF power amplifier? (G4A08)

A. Minimum SWR on the antenna
B. Minimum plate current without exceeding maximum allowable grid current
C. Highest plate voltage while minimizing grid current
D. Desired power output without exceeding maximum allowable plate current

The TUNE and LOAD controls are critical for tuning vacuum-tube RF amplifiers. The TUNE control helps achieve the correct resonant point, seen as a dip in plate current. In contrast, the LOAD control adjusts the output power to prevent damage by keeping the plate current within safe limits.

Duty Cycle

The duty cycle in radio communication refers to the percentage of time a transmitter is actively transmitting versus being idle. Different modes, such as CW (Morse code), SSB (Single Sideband), or digital modes like FT8, have varying duty cycles. Modes with high-duty cycles require the transmitter to be on an output power more frequently, which can place significant stress on the equipment. Understanding the duty cycle of your mode helps avoid overheating and damaging your transmitter, especially during extended transmissions.

Knowing the mode's duty cycle is important because some modes have high-duty cycles that could push your transmitter beyond its average power rating. For example, digital modes like FT8 often have a nearly 100% duty cycle, meaning the transmitter constantly transmits without breaks. This continuous output can cause the transmitter to generate excessive heat and possibly exceed its designed power-handling capability, leading to equipment failure.

Why is it important to know the duty cycle of the mode you are using when transmitting? (G8B08)

A. To aid in tuning your transmitter
B. Some modes have high-duty cycles that could exceed the transmitter's average power rating
C. To allow time for the other station to break in during a transmission
D. To prevent overmodulation

To remember this for the exam, think of the duty cycle as the "workload" on your transmitter. Modes with high-duty cycles require careful monitoring of your equipment's power ratings and cooling system to avoid damage during prolonged use.

Receiver

Receiver performance metrics such as sensitivity, selectivity, and dynamic range significantly impact communication quality. Sensitivity determines a receiver's ability to detect weak signals, which is needed for long-distance communication in challenging conditions. Selectivity allows the receiver to isolate desired signals in crowded bands, rejecting adjacent-channel interference. Dynamic range ensures the receiver can handle both weak and strong signals without distortion, maintaining signal clarity.

A receiver's noise reduction control minimizes unwanted background noise, enhancing the clarity of weak signals. However, as the noise reduction control is increased, more aggressive algorithms are used to suppress noise. While this can help eliminate unwanted sound, it may also result in signal distortion. Reducing noise can inadvertently filter out parts of the desired signal, especially if the noise reduction is set too high.

What happens as a receiver's noise reduction control level is increased? (G4A07)

A. Received signals may become distorted
B. Received frequency may become unstable
C. CW signals may become severely attenuated
D. Received frequency may shift several kHz

Operators should carefully adjust the noise reduction control for optimal use, balancing between reducing background noise and maintaining the signal's integrity. Pushing the control too far may result in garbled or unintelligible audio, even though the background noise is reduced. Hence, finding the right level where signals remain clear without too much distortion is crucial.

What is the purpose of using a receive attenuator? (G4A13)

A. To prevent receiver overload from strong incoming signals
B. To reduce the transmitter power when driving a linear amplifier
C. To reduce power consumption when operating from batteries
D. To reduce excessive audio level on strong signals

A receive attenuator is a device that reduces the strength of incoming signals before they reach the receiver. The primary purpose of using an attenuator is to prevent receiver overload from powerful signals. When a signal is too strong, it can overwhelm the receiver, causing distortion, loss of sensitivity, and difficulty hearing weaker signals. By reducing the signal strength with an attenuator, you can ensure the receiver performs better, especially when strong local signals might interfere with reception.

For example, if you are trying to listen to weak signals from a distant station but a nearby station is transmitting a powerful signal, the receiver might become overloaded. In such cases, activating the attenuator helps manage this situation by reducing the power of the strong signal without significantly affecting weaker signals. This keeps your receiver from being overwhelmed and allows you to hear the signals more clearly.

Receiver Sensitivity and Bandwidth

Receiver sensitivity and bandwidth are crucial factors in radio system performance. Sensitivity determines the receiver's ability to detect weak signals, while bandwidth affects how much of the signal is passed through the receiver without being filtered out. Optimizing these parameters ensures clear reception with minimal noise and interference. Let's explore key aspects related to radio receivers' sensitivity, bandwidth, and filter performance.

Which parameter affects receiver sensitivity? (G7C08)

A. Input amplifier gain
B. Demodulator stage bandwidth
C. Input amplifier noise figure
D. All these choices are correct

Receiver sensitivity is determined by a few critical factors that dictate how well a receiver can pick up weak signals. The input amplifier gain boosts the strength of incoming signals, while the demodulator stage bandwidth must be set to allow the desired signal through without letting in too much noise. The input amplifier's noise figure also plays a big role—lower noise means better sensitivity, allowing the receiver to detect weaker signals without getting overwhelmed by interference.

The bandwidth of a band-pass filter is measured between what two frequencies? (G7C14)

A. Upper and lower half-power
B. Cutoff and rolloff
C. Pole and zero
D. Image and harmonic

The bandwidth of a band-pass filter is the range of frequencies it allows to pass through. It is measured between the upper and lower half-power points, which correspond to the frequencies where the signal is reduced by 3 dB from its maximum level. This ensures that only the desired signal frequencies are passed while filtering out those outside the target range, leading to better signal clarity and less interference.

What term specifies a filter's maximum ability to reject signals outside its passband? (G7C13)

A. Notch depth
B. Rolloff
C. Insertion loss
D. Ultimate rejection

Ultimate rejection refers to a filter's ability to block or attenuate signals outside its passband, ensuring that unwanted signals don't interfere with the desired transmission. A filter with a high ultimate rejection will effectively isolate the desired signal from outside noise and interference, providing clearer communication.

Understanding receiver sensitivity and bandwidth and how they are influenced by gain, noise, and filtering is key to designing and operating efficient radio systems. Properly balancing these factors ensures that weak signals can be received clearly while unwanted interference is minimized.

Filters and Attenuation

In radio communication systems, filters are critical for shaping and controlling signals by allowing certain frequencies to pass through while attenuating others. Attenuation refers to the reduction of signal strength as it passes through a filter or over a distance. Filters, particularly digital signal processing (DSP) filters, offer advanced flexibility, while key terms like insertion loss and cutoff frequency help quantify the performance of these filters. Let's explore some key questions about these concepts.

Which of the following is an advantage of a digital signal processing (DSP) filter compared to an analog filter? (G7C06)

A. A wide range of filter bandwidths and shapes can be created
B. Fewer digital components are required
C. Mixing products are greatly reduced
D. The DSP filter is much more effective at VHF frequencies

A digital signal processing (DSP) filter offers significant advantages over traditional analog filters, especially in terms of flexibility. DSP filters can be reprogrammed to create a wide range of bandwidths and shapes, allowing precise control over which frequencies are passed or attenuated. This versatility makes DSP filters more adaptable for different communication needs, such as reducing noise or improving signal clarity, without physical hardware changes.

What term specifies a filter's attenuation inside its passband? (G7C07)

A. Insertion loss
B. Return loss
C. Q
D. Ultimate rejection

Insertion loss refers to the attenuation of signal strength within a filter's passband, which is the range of frequencies that the filter is designed to pass. Ideally, a filter should allow all signals within its passband to pass through with minimal loss. However, in practice, some attenuation occurs even within this range, and this reduction is quantified as insertion loss. Low insertion loss is crucial for maintaining signal integrity in communication systems.

What is the frequency above which a low-pass filter's output power is less than half the input power? (G7C12)

A. Notch frequency
B. Neper frequency
C. Cutoff frequency
D. Rolloff frequency

The cutoff frequency of a low-pass filter is the point at which the filter begins to significantly attenuate higher frequencies. Beyond this frequency, the output power of the signal is reduced to less than half of the input power. The cutoff frequency is an essential characteristic of any filter because it defines the threshold at which unwanted high-frequency signals are attenuated, helping to maintain clear signal transmission within the desired frequency range.

Filters and attenuation are integral to maintaining signal clarity in radio communications. Digital signal processing (DSP) filters offer incredible flexibility in shaping signals, while important concepts like insertion loss and cutoff frequency help engineers design systems that minimize unwanted noise and maximize signal strength. Understanding these terms will help you optimize and analyze filters for better communication performance.

Noise Blanker

A noise blanker is a feature in many modern radio receivers designed to reduce a receiver's gain during sudden bursts of noise. These bursts of noise create short, high-intensity pulses that can overwhelm the radio signal, making it difficult to hear the desired transmission. The noise blanker works by detecting these pulses and temporarily reducing the receiver's gain, or sensitivity, during the duration of the noise pulse. This process minimizes the impact of the noise, allowing the desired signal to come through more clearly.

How does a noise blanker work? (G4A03)

A. By temporarily increasing received bandwidth
B. By redirecting noise pulses into a filter capacitor
C. By reducing receiver gain during a noise pulse
D. By clipping noise peaks

A noise blanker acts like a filter, automatically lowering the volume when a loud, unwanted noise is detected, ensuring that the signal you want to hear remains intelligible. This feature is especially useful for operators who work in environments with a lot of electronic interference, helping them maintain clear communication.

In summary, as a General, you must now understand the advanced intricacies of transmitters and receivers for effective communication. Ensuring your equipment is optimized for sensitivity, selectivity, and dynamic range will enhance your experience, enabling you to maximize your General license.

Signal Strength Meter

A Signal Strength Meter (S Meter) measures the strength of a received signal in decibels (dB) relative to a standard reference point. It is commonly found on most amateur radio transceivers. It gives an operator a quick visual indication of how strong an incoming signal is. The meter typically ranges from S1 to S9, with each number representing a stronger signal. Beyond S9, additional signal strength can be indicated in decibel increments, like 10 dB over S9. This helps operators gauge the quality of a communication link, making adjustments to ensure clear reception.

What does an S meter measure? (G4D04)

A. Carrier suppression
B. Impedance
C. Received signal strength
D. Transmitter power output

Let's look at some math questions about S Meter.

How does a signal that reads 20 dB over S9 compare to one that reads S9 on a receiver, assuming a properly calibrated S meter? (G4D05)

A. It is 10 times less powerful
B. It is 20 times less powerful
C. It is 20 times more powerful
D. It is 100 times more powerful

When you see a signal on your S meter that reads "20 dB over S9," it means that the signal is significantly stronger than one that simply reads S9. In technical terms, 20 dB over S9 indicates that the signal is 100 times more powerful than an S9 signal.

The relationship between decibels and signal strength is logarithmic, meaning each increase of 10 dB represents a tenfold increase in power. Therefore, 20 dB over S9 means two jumps of 10 dB – one jump from 0 to 10 and then the second jump from 11-20, which results in a signal that is 100 times more powerful ($10 \times 10 = 100$).

Understanding this relationship is important because it shows how much stronger one signal is compared to another. It also helps you adjust your receiver settings properly to enhance weaker signals or manage interference from stronger ones.

Another one.

How much change in signal strength is typically represented by one S unit? (G4D06)

A. 6 dB
B. 12 dB
C. 15 dB
D. 18 dB

One S unit on a signal strength meter (S-meter) typically represents a change of 6 decibels (dB) in signal strength. This means that for every increase of one S unit, the signal strength roughly doubles in power (or decreases by half for a decrease of one S unit). For example, a signal that measures S5 is approximately twice as strong as one at S4.

To answer this question effectively, remember:

- **Key Concept**: 1 S unit = 6 dB change in signal strength.

- **Application**: This applies to the calibration of most S-meters on HF transceivers, where a reading of S9 is the standard for a strong signal.

Knowing this relationship is essential for understanding signal reports and troubleshooting receiver performance.

And one final exam question.

How much must the power output of a transmitter be raised to change the S meter reading on a distant receiver from S8 to S9? (G4D07)

A. Approximately 1.5 times
B. Approximately 2 times
C. Approximately 4 times
D. Approximately 8 times

To change the S-meter reading on a distant receiver from S8 to S9, the transmitter's power output must be increased by a factor of 4.

One S-unit represents a change of 6 decibels (dB) in signal strength, and a 6 dB increase requires the signal power to be quadrupled.

Why Quadruple?

The decibel (dB) scale is logarithmic:

- Every **3 dB** increase doubles the power.

- A **6 dB** increase (one S-unit) is the same as **doubling the power twice:** $2 \times 2 = 4$ **times the power**.

Example:

Let's say your transmitter is currently outputting **25 watts**, and the receiving station's S-meter shows **S8**. To move the reading to **S9**:

- Multiply your current power by 4: **25 watts** $\times$ **4 = 100 watts**.

This relationship applies universally, so you can find the required increase by multiplying it by 4 for any starting power.

Transceiver

A transceiver is a device that combines both a transmitter and a receiver into a single unit, allowing for two-way communication. The name comes from combining the words "transmitter" and "receiver."

The transceiver allows the operator to communicate with other radio operators, both locally and globally, by transmitting their voice, data, or Morse code and receiving signals from others. Many modern transceivers are highly versatile, offering multiple modes of communication such as voice, digital, and even video over radio frequencies.

Notch Filter

Several features help improve communication clarity and ensure smooth operation with external equipment when operating an HF transceiver. One of these features is the notch filter, designed to reduce interference from carriers that may be present within the receiver's passband. A carrier is a continuous signal that can cause unwanted noise or interference, making it challenging to hear weaker signals. The notch filter helps by "notching out" or removing these unwanted carriers, allowing for clearer reception of the desired signal.

> **What is the purpose of the notch filter found on many HF transceivers? (G4A01)**
>
> A. To restrict the transmitter voice bandwidth
> **B. To reduce interference from carriers in the receiver passband**
> C. To eliminate receiver interference from impulse noise sources
> D. To remove interfering splatter generated by signals on adjacent frequencies

Receiver Passband

A receiver passband refers to the range of frequencies that a radio receiver can process and pass through its filters to the output, allowing the operator to hear or decode the signals within that range. It is determined by the receiver's intermediate frequency (IF) filters or digital signal processing (DSP).

- **Purpose**: The passband defines the bandwidth of the receiver, ensuring it includes the desired signal while excluding adjacent signals and noise. For example, a single-sideband signal typically requires a passband of about 2.4–3 kHz, while a CW signal may only need 500 Hz or less.

- **Adjustment**: Many modern receivers allow operators to adjust the passband width and center frequency to optimize reception under different conditions, such as crowded bands or weak signals.

- **Relevance**: The passband directly affects the receiver's ability to isolate desired signals and reject interference, making it a critical factor in effective communication.

Delayed Output

Another feature is the delayed RF output when using an external amplifier. When you key up your transceiver, there needs to be a brief delay before RF energy is sent to the amplifier. This delay allows the amplifier to switch the antenna between the transceiver and its output. Without this delay, RF could be transmitted before the amplifier is ready, potentially causing signal distortion or damage to the equipment.

> **What is the purpose of delaying RF output after activating a transmitter's keying line to an external amplifier? (G4A09)**
>
> A. To prevent key clicks on CW
> B. To prevent transient overmodulation
> **C. To allow time for the amplifier to switch the antenna between the transceiver and the amplifier output**
> D. To allow time for the amplifier power supply to reach operating level

Dual VFO

The dual-VFO (Variable Frequency Oscillator) feature is commonly used for transmitting on one frequency and listening on another. This is particularly useful in situations like split-frequency operation, where one operator transmits on a different frequency than they receive. This allows for more flexible and efficient communication, especially in DX (long-distance) operations.

> **Which of the following is a common use of the dual-VFO feature on a transceiver? (G4A12)**
>
> A. To allow transmitting on two frequencies at once
> B. To permit full duplex operation -- that is, transmitting and receiving at the same time
> **C. To transmit on one frequency and listen on another**
> D. To improve frequency accuracy by allowing variable frequency output (VFO) operation

Speech Processor

A speech processor is a device used in transceivers to enhance voice clarity and loudness by manipulating the audio signal before it's transmitted. It works by compressing the dynamic range of the voice signal, which means it amplifies quieter parts while limiting louder parts, resulting in a more consistent and louder signal. This is particularly useful in weak signal conditions or noisy environments, helping to ensure that the operator's voice stands out clearly over long distances.

What is the purpose of a speech processor in a transceiver? (G4D01)

A. Increase the apparent loudness of transmitted voice signals
B. Increase transmitter bass response for more natural-sounding SSB signals
C. Prevent distortion of voice signals
D. Decrease high-frequency voice output to prevent out-of-band operation

A speech processor in a transceiver is a device designed to enhance the clarity and effectiveness of voice communication by increasing the apparent loudness of transmitted voice signals. This is especially useful in environments where signals may be weak, or there is interference. The speech processor works by amplifying the quieter parts of your speech and compressing the louder parts, ensuring that more of your voice energy is transmitted.

How does a speech processor affect a single sideband phone signal? (G4D02)

A. It increases peak power
B. It increases average power
C. It reduces harmonic distortion
D. It reduces intermodulation distortion

The speech processor improves the average power output of the transmission, making the signal more readable, especially when using modes like single sideband. However, if set incorrectly, it can introduce distortion, affect bandwidth, or degrade the quality of the signal, so proper adjustment is key to its effectiveness.

The main benefit of a speech processor is that it increases the average power of a signal, particularly on single-sideband transmissions. This makes the voice signal stand out better and improves the overall readability of your transmission, which is crucial when dealing with weak signals or noisy band conditions.

What is the effect of an incorrectly adjusted speech processor? (G4D03)

A. Distorted speech
B. Excess intermodulation products
C. Excessive background noise
D. All these choices are correct

However, an incorrectly adjusted speech processor can cause significant issues. The most common problems include distorted speech, where the voice becomes unclear or garbled; excess intermodulation products, which result in unwanted signal mixing that can cause interference; and excessive background noise, making it harder for the receiving station to understand the transmission. These issues reduce the overall effectiveness of communication and may cause interference with other stations, so proper adjustment is needed.

In summary, while a speech processor can significantly enhance your signal, improper settings can lead to degraded transmission quality, affecting clarity and causing unwanted interference. Always ensure it's adjusted correctly for the best performance.

Chapter Summary

Transmitters, receivers, and transceivers are the core components of every ham radio station, transforming theoretical knowledge into practical communication. As a General Class operator, your expanded privileges demand a deeper understanding of these systems to confidently explore new bands and modes. With the right knowledge and tools, you can enhance your station's performance, minimize interference, and enjoy seamless communication across the airwaves. This solid foundation will prepare you for more advanced topics.

Chapter 10

Circuit Components and Design

Basic Electronic Components

ELECTRONIC CIRCUITS ARE COMPOSED of various fundamental components, each serving a specific function. Resistors, capacitors, and inductors are the primary passive components essential for controlling and manipulating electrical signals.

Resistors

Resistors introduce resistance to the flow of current, measured in ohms (Ω). They come in various types, including through-hole and surface-mount, each suited for different applications. Typical uses of resistors include current limiting, voltage division, and feedback circuits. Selecting the proper resistor involves considering its resistance value, power dissipation, and tolerance. These parameters ensure that the resistor can handle the required current without overheating or deviating from its specified value.

Wire-Wound Resistors

While wire-wound resistors are excellent for power handling and precision in certain circuits, they are not suitable for radio frequency (RF) circuits due to their inherent inductance. The wire-wound around the resistor's core can act like a small inductor, introducing inductive reactance at high frequencies. This inductive behavior can interfere with the resistor's performance, making the circuit unpredictable and less efficient in RF applications.

> **Why should wire-wound resistors not be used in RF circuits? (G6A06)**
>
> A. The resistor's tolerance value would not be adequate
> **B. The resistor's inductance could make circuit performance unpredictable**
> C. The resistor could overheat
> D. The resistor's internal capacitance would detune the circuit

As RF circuits rely on precise tuning and performance, using a component that adds unintended inductance could severely affect signal quality or cause the circuit to malfunction.

Power Dissipation

Power dissipation refers to the process by which an electrical or electronic component converts electrical energy into heat or another form of energy that is not useful for the circuit's intended purpose. This happens whenever current flows through a component with resistance, such as a resistor, and energy is lost as heat. Power dissipation is an important concept in circuit design because excessive heat can damage components or reduce the system's efficiency.

Root Mean Square / RMS Voltage

An AC signal's RMS (Root Mean Square) value produces the same power dissipation in a resistor as a DC voltage of the same value. This is because the RMS value represents the effective or "average" voltage or current that would result in the same amount of heat or power loss as a steady DC signal. While the voltage in an AC signal constantly changes, the RMS value gives a way to measure how much actual power the signal delivers, just like a constant DC voltage would.

What value of an AC signal produces the same power dissipation in a resistor as a DC voltage of the same value? (G5B07)

A. The peak-to-peak value
B. The peak value
C. The RMS value
D. The reciprocal of the RMS value

The RMS (Root Mean Square) value expresses the practical or equivalent value of an alternating current or voltage. For AC signals, the current or voltage constantly changes. Hence, the RMS value represents the "average" level of power that would be equivalent to the same amount of power as a direct current signal.

Mathematically, the RMS value is the square root of the mean (average) of the squares of all the instantaneous values of the waveform over one complete cycle. For a sinusoidal AC signal, the RMS value is approximately 0.707 times the signal's peak value. The RMS value is important because it gives a practical measure of the power delivered by an AC signal, making it comparable to DC.

Math Diversion (not needed for the test)

The factor of 0.707 that relates the RMS voltage to the peak voltage comes from the mathematical definition of the Root Mean Square (RMS) value for a sine wave.

Here's a step-by-step explanation:

1. **Understanding RMS**: The RMS value is essentially the square root of the average of the squares of all instantaneous values of a waveform over one complete cycle. For an AC signal, the RMS value gives the equivalent DC value that would produce the same power dissipation in a resistor.

2. **The mathematical relationship**: For a sine wave, the RMS value is derived from integrating the square of the sine function over a complete cycle (from 0 to 2π radians) and then taking the square root of the average of those values. The detailed mathematical derivation shows that the RMS value of a sine wave is equal to the peak value (V_{peak}) divided by the reciprocal of the square root of 2.

3. The formula is:

$$V_{RMS} = \frac{V_{peak}}{\frac{1}{\sqrt{2}}} \approx V_{peak} \times 0.707$$

This factor of 0.707 comes from the fact that the square root of 2, divided by 2 gives us approximately 0.707.

This is why, for a sine wave, the RMS value is approximately 70.7% of the peak value. This relationship is critical to understanding how AC voltages are compared to DC voltages in terms of power dissipation.

Example for Voltage

For an AC voltage with a peak value of V_{peak}:

$$V_{RMS} = V_{peak} \times 0.707$$

So, if the peak voltage is 100 volts, the RMS value is:

$$V_{RMS} = 100 \times 0.707 = 70.7\ volts$$

The RMS value is used because it gives the equivalent voltage or current that would produce the same heating effect (or power) as a DC value in a resistive load.

Ok, let's try some exam questions now.

ELECTRICAL WAVEFORMS

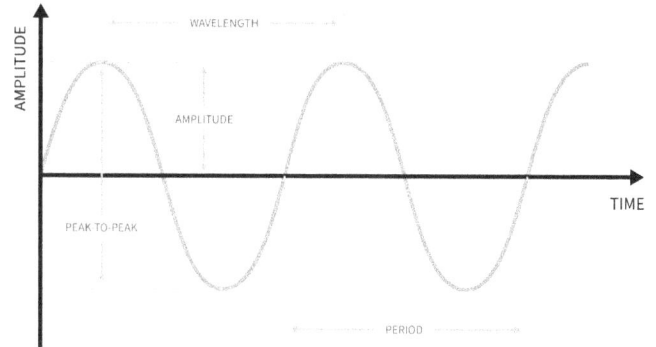

What is the peak-to-peak voltage of a sine wave with an RMS voltage of 120 volts? (G5B08)

A. 84.8 volts
B. 169.7 volts
C. 240.0 volts
D. 339.4 volts

To calculate the peak-to-peak voltage of a sine wave from its RMS voltage, you need to know the relationship between RMS voltage and peak voltage. For a sine wave, the RMS voltage is approximately 0.707 times the peak voltage. The peak-to-peak voltage is then twice the peak voltage.

Here's how you can calculate it:

Step 1: Start with the peak voltage formula.

$$V_{peak} = \frac{V_{RMS}}{0.707}$$

Step 2: Enter the variables from the question. RMS Voltage = 120 volts

$$V_{peak} = \frac{120}{0.707}$$

Step 3: Solve the math.

$$V_{peak} = \frac{120}{0.707} \approx 169.7\ volts$$

Step 4: Find the peak-to-peak voltage using the formula.

$$V_{peak-to-peak} = 2 \times V_{peak}$$

Step 5: Enter the variables.

$$V_{peak-to-peak} = 2 \times 169.7$$

Step 6: Solve the math.

$$V_{peak-to-peak} = 2 \times 169.7 \approx 339.4\ volts$$

So, for a sine wave with an RMS voltage of 120 volts, the peak-to-peak voltage is approximately 339.4 volts.

What is the RMS voltage of a sine wave with a value of 17 volts peak? (G5B09)

A. 8.5 volts
B. 12 volts
C. 24 volts
D. 34 volts

To find the RMS voltage of a sine wave when the peak voltage is known, we start with the same formula:

$$V_{peak} = \frac{V_{RMS}}{0.707}$$

But rearrange it to find RMS Voltage.

$$V_{RMS} = V_{peak} \times 0.707$$

In this case, the peak voltage is 17 volts. Applying the formula and solving.

$$V_{RMS} = 17 \times 0.707 \approx 12.02 \; volts$$

So, the RMS voltage of a sine wave with a peak voltage of 17 volts is approximately 12.02 volts... or 12 volts.

The RMS value gives the equivalent DC voltage that would produce the same power dissipation in a resistor as the AC signal – repetition is key!

RMS Voltage with Power

What is the RMS voltage across a 50-ohm dummy load dissipating 1200 watts? (G5B12)

A. 173 volts
B. 245 volts
C. 346 volts
D. 692 volts

Let's walk through how to find the RMS voltage across a 50-ohm dummy load dissipating 1200 watts based on the foundational formulas we know.

Step 1: Understand the relationship between power, voltage, and resistance = Power Formula

$$Power\ (P) = Current\ (I) \times Voltage\ (E)$$

Step 2: Relate current to voltage and resistance = Ohm's Law

$$Current\ (I) = \frac{Voltage\ (E)}{Resistance\ (R)}$$

Step 3: Substitute 'I' into the power formula

$$Using\ I = \frac{E}{R}\ substitue\ this\ into\ the\ Power\ Formula\ in\ Step\ \#1$$

$$Power\ (P) = \left(\frac{Voltage\ (E)}{Resistance\ (R)}\right) \times Voltage(V)$$

Step 4: Simplify the expression

$$P = \frac{E^2}{R}$$

Or, because we are talking about RMS Voltage, it can be written as

$$P = \frac{V_{RMS}^2}{R}$$

Where:

- P is the power (1200 watts),

- V_{RMS} is the RMS voltage, which we're solving for (and yes you will see V and E both used for voltage)!

- R is the resistance of the load (50 ohms).

Step 5: Rearranging the formula to solve for V_{RMS}

$$V_{RMS} = \sqrt{P \times R}$$

Step 6: Substituting the values and solve

$$V_{RMS} = \sqrt{1200 \times 50} = \sqrt{60,000} \approx 244.9\ volts$$

So, the RMS voltage across the 50-ohm dummy load dissipating 1200 watts is approximately 244.9 or simply 245 volts.

To remember this for the exam, start with the key formulas and recombine based on the variables you are given and what you are solving for.

Peak Envelope Power

Peak Envelope Power (PEP) is the maximum instantaneous power that a transmitter delivers during one cycle of the signal's modulation envelope, typically in an amplitude-modulated (AM), single-sideband (SSB), or other modulated signal. It represents the highest power level the signal reaches as it fluctuates over time due to modulation, which can be voice, data, or other types of information.

PEP is important in radio communication because it measures the highest power output point during transmission, particularly in modulated signals where the power level varies with the modulation (such as voice peaks). In an unmodulated carrier, PEP and average power are the same. Still, in a modulated signal, PEP can be much higher than the average power because it measures the peaks of the signal's envelope.

To sum it up, PEP is used to assess the transmitter's power output during the strongest part of the signal, giving a clear idea of the transmitter's maximum power capability during modulation. This is critical in ensuring transmissions stay within legal power limits while delivering strong and clear signals.

What is the PEP produced by 200 volts peak-to-peak across a 50-ohm dummy load? (G5B06)

A. 1.4 watts
B. 100 watts
C. 353.5 watts
D. 400 watts

Look at this question. What variables are in it? Power (PEP), Voltage (Peak to Peak), and Resistance (Ohm load). Like we did above, we can rearrange the Power Formula and Ohm's Law to get a formula with Power, Voltage and Resistance.

Step 1: Combined Power and Ohm's Law formula (see question above on how we can combine these two to get this formula).

$$P = \frac{V^2}{R}$$

Step 2: Rearrange Peak vs Peak-to-Peak Voltage

$$V_{peak-to-peak} = 2 \times V_{peak}$$

then rearrange to equal

$$V_{peak} = \frac{V_{p-p}}{2}$$

Step 3: Enter values and solve. Peak-to-peak voltage is the total voltage swing from the maximum positive to the maximum negative.

$$V_{peak} = \frac{V_{p-p}}{2}$$

substitue $V_{p-p} = 200\ V$

$$V_{peak} = \frac{200}{2} = 100\ V$$

Step 4: Convert Peak Voltage to RMS Voltage

$$RMS = V_{peak} \times 0.707$$

$$RMS = 100V \times 0.707$$

$$RMS = 70.7V$$

Step 5: Plug the RMS Voltage into the Power Formula from Step 1

$$P_{PEP} = \frac{RMS^2}{R}$$

substitue RMS = 70.7 V and R = 50 Ω

$$P_{PEP} = \frac{70.7^2}{50}$$

Step 6: Solve

$$P_{PEP} = \frac{4,998.5}{50} = 99.97\ W\ or\ simply\ 100\ Watts$$

Answer: The Peak Envelope Power (PEP) produced by 200 volts peak-to-peak across a 50-ohm dummy load is 100 watts.

To remember this for the exam, always convert the peak voltage to RMS before calculating power, as power is usually calculated using RMS voltage in AC circuits.

Do you have this straight? Let's try another one, just in case:

What is the output PEP of 500 volts peak-to-peak across a 50-ohm load? (G5B14)

A. 8.75 watts
B. 625 watts
C. 2500 watts
D. 5000 watts

It is similar to the steps we just walked through in the last question.

Step 1: Combined Power and Ohm's Law formula (see question above on how we can combine these two to get this formula).

$$P = \frac{V^2}{R}$$

Step 2: Rearrange Peak vs Peak-to-Peak Voltage

$$V_{peak-to-peak} = 2 \times V_{peak}$$

this formula rearranged equals

$$V_{peak} = \frac{V_{peak-to-peak}}{2}$$

Step 3: Enter values and solve

substitue $V_{peak-to-peak} = 500\ volts$

$$V_{peak} = \frac{500}{2} = 250\ volts$$

Step 4: Convert the peak voltage to RMS voltage

$$V_{RMS} = V_{peak} \times \frac{1}{\sqrt{2}}$$

substitue $V_{peak} = 250\ volts$

Step 5: Plug the Peak Voltage into the Power Formula from Step 1

the formula for Power when using RMS voltage is:

$$P = \frac{V_{RMS}^2}{R}$$

substitue RMS voltage of 176.75 volts and resistance of 50 ohms

$$P_{PEP} = \frac{176.75^2}{50}$$

Step 6: Solve

$$P_{PEP} = \frac{176.75^2}{50} = \frac{31,240}{50} = 625\ watts$$

The correct output PEP for 500 volts peak-to-peak across a 50-ohm load is 625 watts.

Capacitors

Electrolytic capacitors are known for their high capacitance relative to their physical size. They are ideal for applications requiring large amounts of charge storage in compact spaces. This makes them widely used in power supply circuits, where they help smooth out voltage fluctuations. The ability to store a large charge in a small volume is a defining characteristic of electrolytic capacitors. Still, they are also polarized, meaning they must be connected with the correct polarity in a circuit.

Which of the following is characteristic of an electrolytic capacitor? (G6A04)

A. Tight tolerance
B. Much less leakage than any other type
C. High capacitance for a given volume
D. Inexpensive RF capacitor

Electrolytic capacitors, such as power supplies, are key components in circuits requiring high capacitance in limited space. They are polarized, and connecting them the wrong way can cause failure.

Ceramic Capacitors

Low-voltage ceramic capacitors are known for their comparatively low cost. Due to their affordability, they are commonly used in electronics. These capacitors are ideal for high-frequency applications like decoupling and filtering in circuits. One of the advantages of ceramic capacitors is their stability, which makes them suitable for many different types of circuits, including radio frequency and power supply circuits. While they may not have the high capacitance of electrolytic capacitors, their cost-effectiveness and reliability make them a staple in many applications.

Which of the following is characteristic of low voltage ceramic capacitors? (G6A08)

A. Tight tolerance
B. High stability
C. High capacitance for given volume
D. Comparatively low cost

Low-voltage ceramic capacitors are cost-effective and commonly used in high-frequency applications. They offer stability at a relatively low cost compared to other capacitor types.

Inductor

When an inductor is operated above its self-resonant frequency, its behavior changes from inductive to capacitive. The self-resonant frequency is the point at which the inductance of the coil and the small amount of parasitic capacitance between the windings cancel each other out, effectively making the inductor resonate. Above this frequency, the capacitive effects dominate, and the inductor begins to behave like a capacitor rather than an inductor. This is due to the internal capacitance overtaking the inductance, shifting its overall impedance behavior.

What happens when an inductor is operated above its self-resonant frequency? (G6A11)

A. Its reactance increases
B. Harmonics are generated
C. It becomes capacitive
D. Catastrophic failure is likely

Inductors are typically designed to work below their self-resonant frequency. They lose their intended inductive properties when pushed beyond this frequency and act more like capacitors.

Diode

A diode is an electronic component that allows current to flow in one direction only, functioning as a one-way valve for electricity. It has two terminals: the anode (positive) and the cathode (negative). The diode allows current to pass through when the voltage applied to the anode is more positive than the cathode (a condition known as forward bias). However, the diode blocks current flow when the voltage is reversed (reverse bias).

Diodes are used in a wide variety of applications, including rectifying alternating current (AC) to direct current (DC), protecting circuits from voltage spikes, and in devices like LEDs (Light Emitting Diodes) for light emission.

Forward Biased

An LED (Light-Emitting Diode) emits light when it is forward-biased, meaning current flows from the anode (positive) to the cathode (negative). In this state, electrons recombine with holes in the LED's semiconductor material, releasing energy in the form of light. Forward biasing creates the necessary conditions for this electron-hole recombination, which is key to the light emission process in LEDs. Unlike regular diodes, LEDs are specially designed to emit visible light or infrared light when forward-biased.

How is an LED biased when emitting light? (G6B08)

A. In the tunnel-effect region
B. At the Zener voltage
C. Reverse biased
D. Forward biased

Forward biasing is essential for an LED to function and emit light. It allows current to flow through the device and enables the release of light energy. This concept is critical in understanding how LEDs operate in various applications, from indicators to communication systems.

LED vs LCD

Let's take a quick diversion here. Modern ham radio equipment often features either liquid crystal displays (LCDs) or light-emitting diode (LED) displays, each with its advantages depending on the operating environment.

LCDs are commonly found in handheld and mobile transceivers due to their ability to provide clear, high-contrast visuals in bright lighting (ambient) conditions, making them ideal for outdoor use.

On the other hand, LED displays are known for their bright and vivid colors, making them excellent for indoor or dimly lit environments. Understanding these differences helps operators select the right display type for their specific needs, ensuring clear readability during operation.

How does a liquid crystal display compare to an LED display? (G6B09)

A. Higher contrast in high ambient lighting
B. Wider dynamic range
C. Higher power consumption
D. Shorter lifetime

 Think of it like reading a book versus looking at a phone screen outside—the book's print remains clear, while the phone's screen might glare or fade. This is why many portable and outdoor ham radios use LCD screens, as they provide higher contrast in high ambient lighting and ensure readability even in challenging conditions.

Germanium Diode

Germanium diodes have a lower forward threshold voltage than silicon diodes, approximately 0.3 volts. This lower voltage means that germanium diodes can begin conducting current at lower voltages than silicon diodes, making them useful in low-voltage applications.

What is the approximate forward threshold voltage of a germanium diode? (G6A03)

A. 0.1 volt
B. 0.3 volts
C. 0.7 volts
D. 1.0 volts

Germanium diodes require about 0.3 volts to start conducting, making them suitable for low-voltage applications.

Silicon Junction Diode

Unlike germanium diodes, silicon junction diodes have a higher forward threshold voltage, typically around 0.7 volts. This means that in order for current to flow through a silicon diode, the applied voltage must be at least 0.7 volts – which is double germanium's. Silicon diodes are commonly used in general-purpose rectifiers and switching circuits.

What is the approximate forward threshold voltage of a silicon junction diode? (G6A05)

A. 0.1 volt
B. 0.3 volts
C. 0.7 volts
D. 1.0 volts

Silicon junction diodes have a forward threshold voltage of around 0.7 volts. They are widely used in various electronic applications such as rectification and switching.

Transistors

Bipolar Transistors

A bipolar transistor used as a switch operates in two distinct operating points: saturation and cutoff. In the saturation region, the transistor is fully "on," meaning current flows freely from the collector to the emitter, allowing the transistor to conduct electricity. This state is similar to a closed switch, where the transistor provides a low-resistance path for current. On the other hand, in the cutoff region, the transistor is fully "off," meaning no current flows from the collector to the emitter, acting as an open switch with high resistance. The transistor can effectively control whether current flows by switching between these two states.

What are the operating points for a bipolar transistor used as a switch? (G6A07)

A. Saturation and cutoff
B. The active region (between cutoff and saturation)
C. Peak and valley current points
D. Enhancement and depletion modes

When used as switches, bipolar transistors operate in either saturation (fully on) or cutoff (fully off), enabling them to control current flow in circuits, similar to turning a switch on or off.

MOSFET

A MOSFET (Metal-Oxide-Semiconductor Field-Effect Transistor) is a type of transistor. Its construction involves a gate separated from the channel (the path through which current flows) by a thin insulating layer, typically made of silicon dioxide. This insulating layer allows the MOSFET to control the flow of current between the source and drain without a direct electrical connection, making it highly efficient and minimizing power loss. The gate voltage regulates the conductivity of the channel, allowing the MOSFET to act as a switch or amplifier.

Which of the following describes MOSFET construction? (G6A09)

A. The gate is formed by a back-biased junction
B. The gate is separated from the channel by a thin insulating layer
C. The source is separated from the drain by a thin insulating layer
D. The source is formed by depositing metal on silicon

MOSFETs are important to modern electronics because they can control large currents with minimal power. The insulated gate is a defining feature that enables them to efficiently switch and amplify signals in devices ranging from computers to radios.

Vacuum Tube

A vacuum tube, also known as a thermionic valve, is an electronic device used to control the flow of electrical current in a sealed glass or metal enclosure with no air inside. It works by heating a cathode (negative electrode) to release electrons into the vacuum, which are then attracted to the anode (positive electrode), creating a current. A control grid between the cathode and anode allows for regulating this flow, enabling the vacuum tube to amplify or modulate signals. Common types include diodes (converting AC to DC) and triodes (amplifying signals). Although largely replaced by transistors, vacuum tubes are still used in high-power RF transmitters, audiophile-grade audio equipment, and some military and scientific applications due to their reliability in extreme conditions and unique performance characteristics.

Control Grid

Which element of a vacuum tube regulates the flow of electrons between cathode and plate? (G6A10)

A. Control grid
B. Suppressor grid
C. Screen grid
D. Trigger electrode

In a vacuum tube, the control grid is the component responsible for regulating the flow of electrons between the cathode (which emits electrons) and the plate (which collects them). By applying a varying voltage to the control grid, you can control the amount of current that flows through the tube. This is the fundamental principle behind amplification in vacuum tubes. Essentially, the control grid allows small changes in voltage to create larger changes in current, enabling the tube to amplify signals effectively.

Vacuum Tubes

Screen Grid

What is the primary purpose of a screen grid in a vacuum tube? (G6A12)

A. To reduce grid-to-plate capacitance
B. To increase efficiency
C. To increase the control grid resistance
D. To decrease plate resistance

The screen grid serves a different function. It is placed between the control grid and the plate, and its primary purpose is to reduce grid-to-plate capacitance. This helps to stabilize the tube's operation at higher frequencies by preventing unwanted feedback and oscillations. By shielding the control grid from the plate, the screen grid ensures that the tube can operate more efficiently in RF applications.

The control grid manages the electron flow, making the tube act as an amplifier. The screen grid helps maintain stability by minimizing interference between the control grid and the plate, particularly in high-frequency applications.

Connectors

Connectors link different radio system components, such as transceivers, antennas, microphones, and power supplies. They ensure that signals are efficiently transmitted without interference, loss, or damage to the equipment.

SMA Connector

What is an SMA connector? (G6B11)

A. A type-S to type-M adaptor
B. A small threaded connector suitable for signals up to several GHz
C. A connector designed for serial multiple access signals
D. A type of push-on connector intended for high-voltage applications

A SMA connector (SubMiniature version A) is a small, threaded RF (Radio Frequency) connector commonly used for frequencies up to several GHz. It's designed for tight spaces and is often used in applications such as wireless communication systems, antennas, and test equipment. SMA connectors are known for their precision and durability, making them suitable for high-frequency use in ham radio and other RF applications.

For exam purposes, remember that an SMA connector is suitable for signals up to several GHz, making it ideal for high-frequency use. This is why they are popular for handheld transceivers, wireless devices, and even GPS antennas.

Connectors

Type N Connector

Which of the following describes a type N connector? (G6B07)

A. A moisture-resistant RF connector useful to 10 GHz
B. A small bayonet connector used for data circuits
C. A low noise figure VHF connector
D. A nickel-plated version of the PL-259

A Type N connector is a threaded RF connector designed to operate in frequencies up to 10 GHz. Its key feature is that it is moisture-resistant, making it suitable for outdoor applications where weatherproofing is essential. Type N connectors provide excellent low-loss performance, which is why they are commonly used in microwave and high-frequency applications like base stations and antenna setups for UHF and microwave bands.

The points to recall for the exam are that a Type N connector is moisture-resistant and can be used in frequencies as high as 10 GHz. Its durability makes it a preferred choice for outdoor installations and higher-frequency operations.

BNC Connector

The BNC connector (Bayonet Neill-Concelman) is a commonly used RF connector, but its performance is limited at higher frequencies. The typical upper-frequency limit for low SWR (Standing Wave Ratio) operation for 50-ohm BNC connectors is 4 GHz. While BNC connectors are popular for applications like oscilloscopes, network analyzers, and video transmission, they may cause higher losses or reflections at frequencies beyond 4 GHz.

What is a typical upper frequency limit for low SWR operation of 50-ohm BNC connectors? (G6B04)

A. 50 MHz
B. 500 MHz
C. 4 GHz
D. 40 GHz

The 4 GHz frequency is in the microwave spectrum with a wavelength of 7.5 cm.

For the exam, remember that BNC connectors are limited to low SWR operation up to 4 GHz, and beyond this point, they may not perform as well in maintaining signal integrity.

RCA Phono Connector

Which of these connector types is commonly used for low frequency or dc signal connections to a transceiver? (G6B12)

A. PL-259
B. BNC
C. RCA Phono
D. Type N

The RCA Phono connector is commonly used for low-frequency or DC signal connections to a transceiver. This type of connector, recognizable by its round shape with a central pin and a ring around it, is often seen in consumer electronics like audio and video equipment. In ham radio and other electronic systems, RCA connectors are typically used for connecting low-frequency audio signals or control signals, such as the connection between an audio output and an amplifier or a signal input into a transceiver.

RCA Phono connectors are widely used due to their simplicity, ease of use, and availability. They are reliable for low-frequency signals, making them a suitable choice for applications that don't require handling high-power or high-frequency RF signals. These connectors have been a standard in electronics for decades, primarily because they offer solid connections for various audio and control applications in transceiver setups.

Ferrite Core

A ferrite core is a type of magnetic core made from a ceramic compound composed of iron oxide mixed with metallic elements such as zinc, manganese, or nickel. These cores suppress high-frequency noise in electrical circuits, particularly transformers, inductors, and other electromagnetic devices. The high magnetic permeability of ferrite cores allows them to store magnetic energy effectively while minimizing electrical losses at high frequencies. In radio communication equipment, ferrite cores are often used to improve the performance of antennas and transformers.

Ferrite Core Performance at Different Frequencies

Ferrite cores are widely used in electronics and radio frequency circuits to manage electromagnetic interference (EMI) and enhance the performance of inductors and transformers. The performance of a ferrite core is largely determined by the composition, or "mix," of materials used in its construction. Different mixtures of iron oxide and other metal oxides like zinc or manganese provide various electrical and magnetic properties, making specific ferrite cores more suited for particular frequency ranges. For instance, some ferrite mixes work better at high frequencies. In contrast, others are more effective at lower frequencies, allowing designers to choose the suitable core for optimal suppression of unwanted signals.

What determines the performance of a ferrite core at different frequencies? (G6B01)

A. Its conductivity
B. Its thickness
C. The composition, or "mix," of materials used
D. The ratio of outer diameter to inner diameter

Ferrite Core Toroidal Inductors

A ferrite core toroidal inductor offers several advantages in RF applications. One key benefit is that it allows for the creation of large inductance values while occupying relatively little space. The design of the toroid helps confine most of the magnetic field within the core itself, reducing electromagnetic interference (EMI) with nearby components. Additionally, the magnetic properties of the ferrite material can be optimized for a specific frequency range, making these inductors highly efficient and effective in RF circuits. This leads to better energy transfer and reduced losses.

What is an advantage of using a ferrite core toroidal inductor? (G6B05)

A. Large values of inductance may be obtained
B. The magnetic properties of the core may be optimized for a specific range of frequencies
C. Most of the magnetic field is contained in the core
D. All these choices are correct

Ferrite Bead

A ferrite bead, on the other hand, is a specific type of ferrite core that is shaped like a small cylindrical bead. It is placed around cables to suppress high-frequency electromagnetic interference (EMI) by acting as a passive filter. Ferrite beads are often found on USB cables, power cords, and signal lines, where they help to reduce noise and prevent interference from affecting the performance of connected devices. Ferrite beads work by increasing the impedance at high frequencies, dissipating unwanted energy such as heat, and thus blocking or filtering out noise.

How does a ferrite bead or core reduce common-mode RF current on the shield of a coaxial cable? (G6B10)

A. By creating an impedance in the current's path
B. It converts common-mode current to differential mode current
C. By creating an out-of-phase current to cancel the common-mode current
D. Ferrites expel magnetic fields

Both ferrite cores and beads help minimize interference in electronic systems and ensure smooth operation across a wide range of frequencies.

Integrated Circuit

An integrated circuit (IC) is a small electronic device made up of a collection of components such as transistors, resistors, diodes, and capacitors fabricated together on a single piece of semiconductor material, usually silicon. These components are connected to perform specific functions like amplification, switching, or digital data processing.

ICs are vital to modern electronics because they allow many components to be packed into a small chip, drastically reducing the size and cost of circuits while improving performance. Integrated circuits are used in a wide range of devices, including computers, smartphones, radios, and televisions.

The key advantage of ICs is that they integrate multiple components into a single chip, allowing for complex functionality in a compact and efficient form. This makes modern technology more powerful, efficient, and accessible.

Monolithic Microwave Integrated Circuit (MMIC)

What is meant by the term MMIC? (G6B02)

A. Multi-Mode Integrated Circuit
B. Monolithic Microwave Integrated Circuit
C. Metal Monolayer Integrated Circuit
D. Mode Modulated Integrated Circuit

An MMIC (Monolithic Microwave Integrated Circuit) is a type of integrated circuit specifically designed to operate at microwave frequencies, typically in the GHz range. Unlike traditional ICs, which work at lower frequencies, MMICs are used in applications like radar systems, satellite communication, and high-frequency wireless systems. The term "monolithic" refers to the fact that the circuit components, such as transistors and capacitors, are fabricated on a single semiconductor chip. The compactness of MMICs allows for efficient high-frequency performance, making them ideal for communication systems that require small, high-speed components.

CMOS vs TTL Integrated Circuits

Two of the most common families of ICs are CMOS (Complementary Metal-Oxide-Semiconductor) and TTL (Transistor-Transistor Logic). Each has unique characteristics, making one more suitable than the other depending on the application.

Which of the following is an advantage of CMOS integrated circuits compared to TTL integrated circuits? (G6B03)

A. Low power consumption
B. High power handling capability
C. Better suited for RF amplification
D. Better suited for power supply regulation

One significant advantage of CMOS over TTL is its low power consumption. CMOS circuits consume much less power at rest, as power is mainly drawn during switching, making them ideal for battery-operated devices. In contrast, TTL circuits consume more power continuously. Additionally, CMOS circuits tend to be more resistant to noise, which is another reason they are widely used in modern electronics like microcontrollers, smartphones, and digital logic devices.

Digital Circuits and Logic Gates

Digital circuits form the backbone of modern electronics, including ham radios. These circuits operate using logic gates, which process binary inputs to produce specific outputs based on set rules. Logic gates and binary counters are essential in controlling digital operations. At the same time, shift registers play a significant role in data handling and transmission. This section will help you understand how digital components work and why they're critical in ensuring smooth operations within your radio systems.

Which of the following describes the function of a two-input AND gate? (G7B03)

A. Output is high when either or both inputs are low
B. Output is high only when both inputs are high
C. Output is low when either or both inputs are high
D. Output is low only when both inputs are high

An AND gate is a basic digital logic gate that <u>outputs only a "high" (or 1) signal when all its inputs are high.</u> In the case of a two-input AND gate, this means that both inputs must be 1 for the output to also be 1. If either input is 0, the output will be 0. This type of gate is commonly used in circuits where multiple conditions must be met simultaneously, making it a vital building block in digital logic.

How many states does a 3-bit binary counter have? (G7B05)

A. 3
B. 6
C. 8
D. 16

A 3-bit binary counter has <u>eight distinct states,</u> counting from 000 to 111 in binary, which translates to decimal numbers 0 through 7. The number of states in a binary counter is determined by the number of bits it contains, and each additional bit doubles the number of possible states. These counters are vital in timing, counting, and control applications in digital electronics, including radios.

What is a shift register? (G7B06)

A. A clocked array of circuits that passes data in steps along the array
B. An array of operational amplifiers used for tri-state arithmetic operations
C. A digital mixer
D. An analog mixer

<u>A shift register is a series of flip-flops connected in sequence that passes binary data from one flip-flop to the next at each clock cycle.</u> It stores and transfers data in digital circuits. The clock pulses drive the transfer of data, and shift registers are often used in data handling applications like serial-to-parallel data conversion or for managing data transmission in communication systems.

Understanding how digital circuits, logic gates, and components like binary counters and shift registers work is key for mastering the digital aspects of modern radio systems. These elements enable the processing and handling of binary data, which is the foundation of all digital communication. By learning these basic principles, operators can ensure better control and efficiency in their radio operations.

Amplifiers And Oscillators

Amplifier Classes and Efficiency

Amplifiers are used by radios to increase the power of signals to suitable levels for transmission or processing. The different classes of amplifiers—Class A, B, AB, and C—vary in efficiency and performance, depending on the application. This section covers concepts such as neutralization to prevent unwanted oscillations and the comparative efficiencies of different amplifier classes.

What is the purpose of neutralizing an amplifier? (G7B01)

A. To limit the modulation index
B. To eliminate self-oscillations
C. To cut off the final amplifier during standby periods
D. To keep the carrier on frequency

Neutralizing an amplifier ensures that unwanted feedback doesn't cause the amplifier to generate oscillations on its own, known as self-oscillations. This is particularly important in RF amplifiers, where unintended oscillations can lead to poor signal quality or interference. Neutralization techniques adjust the amplifier's internal capacitances to counteract the feedback that could otherwise cause oscillations.

Which of these classes of amplifiers has the highest efficiency? (G7B02)

A. Class A
B. Class B
C. Class AB
D. Class C

Class C amplifiers are the most efficient, often reaching efficiencies above 70%. However, because of their nonlinear operation, they are not suitable for all modes of transmission. Class C amplifiers are typically used for frequency-modulated (FM) signals and other signals where signal distortion is acceptable since they amplify only a portion of the input signal.

For which of the following modes is a Class C power stage appropriate for amplifying a modulated signal? (G7B11)

A. SSB
B. FM
C. AM
D. All these choices are correct

Class C amplifiers are ideal for FM transmissions because their output is not directly related to the amplitude of the input signal. Frequency modulation relies on varying frequencies, so a Class C amplifier's tendency to distort amplitude signals does not affect the overall transmission quality.

In a Class A amplifier, what percentage of the time does the amplifying device conduct? (G7B04)

A. 100%
B. More than 50% but less than 100%
C. 50%
D. Less than 50%

A Class A amplifier conducts 100% of the time, meaning the amplifying device is always active during both halves of the signal. This makes Class A amplifiers highly linear and ideal for high-fidelity amplification. However, their continuous operation results in lower efficiency, typically around 25-30%, because they consume power even when no signal is being amplified.

How is the efficiency of an RF power amplifier determined? (G7B08)

A. Divide the DC input power by the DC output power
B. Divide the RF output power by the DC input power
C. Multiply the RF input power by the reciprocal of the RF output power
D. Add the RF input power to the DC output power

The efficiency of an RF amplifier is calculated by dividing the RF power output by the DC input power. This ratio indicates how effectively the amplifier converts the input energy into useful output signal power. Higher efficiency reduces heat dissipation and energy waste, particularly important in high-power applications.

Understanding amplifier classes, their efficiency, and their applications in radio is fundamental for selecting the right type for different transmission needs. Class C amplifiers offer high efficiency but are limited to specific uses like FM. In contrast, Class A amplifiers provide high linearity but at the cost of efficiency. Properly neutralizing an amplifier prevents unwanted oscillations, improving overall performance. In summary, the choice of amplifier type depends on the specific requirements of the radio signal and application.

Amplifier Characteristics

Amplifiers play a vital role in ensuring that the transmitted or processed signal is strong enough for effective communication. One key characteristic to consider when choosing an amplifier is whether it is linear or nonlinear. Linear amplifiers maintain the integrity of the signal's waveform throughout amplification, making them ideal for modes of communication where accuracy and signal fidelity are critical. This section will cover linear amplification and its significance in radio operations.

Which of the following describes a linear amplifier? (G7B10)

A. Any RF power amplifier used in conjunction with an amateur transceiver
B. An amplifier in which the output preserves the input waveform
C. A Class C high efficiency amplifier
D. An amplifier used as a frequency multiplier

A linear amplifier accurately amplifies an input signal without altering its waveform. This is crucial in applications such as SSB and AM, where the shape of the signal conveys essential information. A linear amplifier ensures that the amplified output is a scaled version of the input, preventing distortion that would degrade the signal quality. While linear amplifiers are less efficient than Class C amplifiers, their precision is critical for these communication modes.

Linear amplifiers ensure that signals are amplified without distortion, preserving the integrity of the waveform. This makes them indispensable for modes like AM and SSB, where even slight distortions can lead to a loss of important signal information. Radio operators can use the appropriate amplifier to ensure high-fidelity transmissions and efficient communications.

Oscillators and Frequency Determination

Oscillators are components in radio frequency circuits and generate the signal frequencies needed for transmission. These circuits convert DC into a periodic signal, often a sine wave, at a precise frequency. The frequency at which an oscillator operates is critical for maintaining clear communication. Key components, such as filters and amplifiers, stabilize and shape the signals. This section will cover the basic components of oscillators and the factors that determine their frequency.

Which of the following are basic components of a sine wave oscillator? (G7B07)

A. An amplifier and a divider
B. A frequency multiplier and a mixer
C. A circulator and a filter operating in a feed-forward loop
D. A filter and an amplifier operating in a feedback loop

A sine wave oscillator generates continuous waveforms, such as the sine wave. It relies on two key components: a filter and an amplifier. These two elements are arranged in a feedback loop, where the filter selects the correct frequency, and the amplifier increases the signal strength. The continuous feedback ensures that the oscillations remain stable and uninterrupted. Sine wave oscillators are used in various applications, including transmitters, signal generators, and clock circuits.

What determines the frequency of an LC oscillator? (G7B09)

A. The number of stages in the counter
B. The number of stages in the divider
C. The inductance and capacitance in the tank circuit
D. The time delay of the lag circuit

The frequency of an LC oscillator is determined by the inductance (L) and capacitance (C) values in its tank circuit. The tank circuit stores energy and sets the oscillation frequency. The larger the inductance or capacitance, the lower the frequency of oscillation. It helps operators fine-tune their transmitters to the desired frequency band.

Oscillators are the heart of signal generation in RF circuits, and their frequency is determined by key components such as inductors and capacitors in the tank circuit. Understanding the relationship between these components is crucial for setting precise frequencies and maintaining stable communication.

Direct Digital Synthesizer

Which of the following is characteristic of a direct digital synthesizer (DDS)? (G7C05)

A. Extremely narrow tuning range
B. Relatively high-power output
C. Pure sine wave output
D. Variable output frequency with the stability of a crystal oscillator

A direct digital synthesizer (DDS) allows for precise and stable frequency generation. Digital techniques can vary the output frequency while maintaining the stability of a crystal oscillator, making DDS useful in applications requiring accurate and stable frequency control, such as modern radio transceivers. DDS is preferred because of its ability to rapidly and precisely generate frequencies across a wide range, making it ideal for frequency-agile systems.

Operational Amplifiers

What kind of device is an integrated circuit operational amplifier? (G6B06)

A. Digital
B. MMIC
C. Programmable Logic
D. Analog

An operational amplifier (op-amp) is a versatile analog device commonly used in signal processing, filtering, or mathematical operations such as addition and integration. The op-amp is designed to amplify the difference in voltage between two input terminals and is highly useful in analog circuits. <u>Being an analog device</u> means that it processes continuous signals, making it crucial for applications like audio processing, control systems, and analog-to-digital conversion. Its high gain and flexibility make it a fundamental building block in a variety of electronic circuits.

Chapter Summary

This chapter has provided a detailed exploration of the critical components that form the foundation of circuit design and functionality in amateur radio. From understanding the behavior of resistors and capacitors to grasping the role of transistors, amplifiers, oscillators, and more, you've gained insight into the key elements that bring radio circuits to life.

Whether fine-tuning frequencies with LC oscillators, selecting the right diodes for rectification or choosing between CMOS and TTL integrated circuits, these principles are central to building and maintaining reliable radio systems. Armed with this knowledge, you're better prepared to troubleshoot, innovate, and optimize your equipment, empowering you to grow as a confident and capable amateur radio operator.

Chapter 11

Circuit Diagrams

Circuit diagrams are the blueprints of electronic circuits, offering a visual representation of how components are connected. They simplify complex circuits, making it easier to understand their functionality. For ham radio operators, these diagrams are indispensable tools for both designing new circuits and troubleshooting existing ones. By laying out each component and connection, circuit diagrams help you visualize the flow of current and the interaction between various elements. This clarity is crucial for diagnosing issues and ensuring that your designs function as intended. Moreover, circuit diagrams are a universal language for sharing designs with others, facilitating collaboration and innovation within the amateur radio community.

Reading and interpreting circuit diagrams involves understanding the symbols and layout that represent different components and their connections. You need to understand series and parallel connections. Components in series share a single path for the current flow, while parallel components provide multiple paths. Schematic notations and labels offer additional information, such as component values and specifications. These details are often annotated near the symbols to give clarity and context.

Using standard conventions, creating circuit diagrams involves translating your design ideas into a visual format. Circuit design software like KiCad and EasyEDA can be invaluable tools for this task. These programs offer libraries of component symbols and allow you to draw connections quickly. You can also simulate your designs to test their functionality before building them. For simpler circuits, hand-drawing techniques can be equally effective. Use a pencil and graph paper to sketch out your design, ensuring that each component is clearly labeled and that connections are neat and organized. Best practices for precise and accurate diagrams include keeping lines straight and avoiding unnecessary crossings, which can make the diagram confusing. Ensure that power and ground connections and group-related components are clearly marked to make the diagram easier to follow.

Schematic Symbols

This section will explore the common symbols used in schematics, their meanings, and how they help communicate complex electrical designs.

Exam Schematic

Here is the schematic that will be on the exam. Let's review the labeled components.

Figure G7-1

Let's walk through the labeled components in the schematic, which correspond to the numbered symbols one through eleven:

1. **Field-Effect Transistor (FET)**: A FET is a transistor used to control the flow of current with an electric field. It's widely used in radios for signal amplification due to its high input impedance, which allows for efficient control with little power. Fun Fact: FETs are used in radio because they can operate at very high frequencies, making them ideal for RF amplification.

2. **NPN Junction Transistor**: This is a Bipolar Junction Transistor (BJT) that allows current to flow when a small positive voltage is applied to its base. It amplifies weak RF or audio signals into stronger ones that can be transmitted or further processed. Interesting Fact: NPN transistors can switch on and off incredibly fast, making them key in digital signal processing.

3. **Diode/ Rectifier**: A diode is a component that allows current to flow in only one direction, making it helpful in converting alternating current (AC) to direct current (DC). Rectifiers, which are configurations of diodes, are used in powering circuits by turning AC into the DC that radios need. Diodes also protect circuits from voltage spikes and are important in voltage regulation.

4. **Varactor Diode**: A varactor diode, also known as a varicap diode, controls capacitance in a circuit. It acts like a variable capacitor, where the capacitance changes based on its voltage. Varactor diodes are often used in frequency tuning, allowing the radio to select different channels easily. They are key in voltage-controlled oscillators and other circuits requiring frequency adjustment.

5. **Zener Diode**: The Zener diode's ability to maintain a fixed voltage in reverse bias makes it a go-to component for voltage regulation in radio circuits. Zener diodes are commonly used in noise reduction circuits, which helps improve radio signal quality.

6. **Solid Core Transformer**: Transformers are used to step up or down the voltage in power supplies or radio circuits. The solid core transfomer transfers energy between circuits through electromagnetic induction. In radio transmitters, transformers help match impedance between stages, ensuring efficient signal transfer.

7. **Tapped Inductor**: Tapped inductors are coils with an extra connection (tap), providing adjustable inductance. They're used for tuning and filtering. Tapped inductors are integral in antenna tuning, which helps radios receive clearer signals by optimizing the antenna's electrical length.

8. **Polarized Capacitor**: A polarized capacitor has a positive and negative terminal, meaning it must be connected in the correct direction in a circuit. Commonly used in power supply circuits, polarized capacitors store and release electrical energy. They are typically used in applications requiring larger capacitance, like filtering or smoothing signals. Connecting them with the wrong polarity can damage the capacitor or the circuit.

9. **Fixed Resistor**: A fixed resistor is a component that resists the flow of electrical current. Its resistance value is set and cannot be changed. It's used to control the amount of current in a circuit, helping to protect components or regulate voltage levels. Fixed resistors are widely used in all types of circuits for their simplicity and reliability in maintaining a consistent resistance value.

10. **Polarized Capacitor**: The same as #8.

11. **Potentiometer**: A potentiometer is a variable resistor that allows you to adjust the resistance within a circuit manually. It has three terminals: two connected to a resistive element and one connected to a movable wiper. Turning the wiper allows you to change the resistance between the terminals, effectively controlling voltage or current flow. Potentiometers are commonly used for volume controls in audio devices and for adjusting levels in various electronic circuits.

You have all those memorized, right? Nope? Well, don't worry too much. This is another one of those times when we use a little bit of exam strategy. While you will learn all the symbols and components over time, let's focus on the five you could see on the exam.

Exam Symbols

Which symbol in Figure G7-1 represents a field effect transistor? (G7A09)

A. Symbol 2
B. Symbol 5
C. Symbol 1
D. Symbol 4

A FET is a transistor that controls current flow using an electric field, making it ideal for amplifiers or switches due to its high input impedance and low power consumption.

Memory Trick: Think of the FET symbol as a **"gatekeeper" with a river flowing underneath.** The "gate" (the control pin) regulates the flow of electrons (the river), just like how the FET controls current. The straight line (drain) and arrow (source) show the flow direction.

Visual Cue: Picture a little toll booth on a bridge – only those allowed by the gatekeeper (the gate pin) can pass!

Which symbol in Figure G7-1 represents a Zener diode? (G7A10)

A. Symbol 4
B. Symbol 1
C. Symbol 11
D. Symbol 5

A Zener diode allows current to flow in the reverse direction when a specific voltage (the Zener voltage) is reached, making it essential for voltage regulation and protection circuits.

Memory Trick: The Zener diode's symbol looks like a regular diode but with a "zippy" zigzag at the tip. Think of it as a **"lightning bolt shield"** that activates when voltage gets too high, protecting your circuit.

Visual Cue: Imagine a superhero shield with a lightning bolt on it, standing guard to stop voltage surges!

Which symbol in Figure G7-1 represents an NPN junction transistor? (G7A11)

A. Symbol 1
B. Symbol 2
C. Symbol 7
D. Symbol 11

An NPN transistor is a type of bipolar junction transistor (BJT) that amplifies or switches electrical signals by controlling the current flow from the collector to the emitter, with a small current applied to the base.

Memory Trick: The NPN transistor has an arrow pointing **"Not Pointing iNward"** (NPN). This outward arrow represents the emitter pushing current out.

Visual Cue: Picture a rocket blasting off, with the arrow showing the current "blasting out" of the emitter.

Which symbol in Figure G7-1 represents a solid core transformer? (G7A12)

A. Symbol 4
B. Symbol 7
C. Symbol 6
D. Symbol 1

A transformer with a solid (non-laminated) core, typically used in applications where high-frequency signals need efficient coupling or where size and weight are critical.

Memory Trick: The transformer symbol has two coils (representing its windings) with a core between them. Think of it as a **"teleporter portal"** where electricity enters one side (primary) and exits the other (secondary), transformed!

Visual Cue: Imagine the core as the magic doorway linking the two coils like in a sci-fi movie.

Which symbol in Figure G7-1 represents a tapped inductor? (G7A13)

A. Symbol 7
B. Symbol 11
C. Symbol 6
D. Symbol 1

An inductor with one or more intermediate connections (taps) along its winding, enabling it to be used for impedance matching, filtering, or voltage step-up/step-down applications.

Memory Trick: The tapped inductor has extra lines (taps) poking out of the coil, like branches growing off a tree. Think of it as a **"tree of options,"** letting you pick different voltages or signals.

Visual Cue: Picture a winding staircase with little "landings" along the way where you can stop and "tap" into the energy.

Creating and interpreting circuit diagrams is fundamental for any ham radio operator. Whether you are designing a new project or troubleshooting an existing one, these diagrams offer invaluable insights into the workings of electronic circuits. They help you visualize connections, simplify complex designs, and communicate your ideas effectively.

Troubleshooting Common Circuit Issues

In electronic circuits, encountering issues is inevitable. Understanding typical problems and their symptoms is crucial for swift and effective troubleshooting.

A systematic approach to troubleshooting starts with a visual inspection. Begin by examining the circuit for obvious signs of damage, such as burnt components, broken wires, or loose connections. Look for any discoloration or physical deformities that might indicate overheating or stress. Continuity testing with a multimeter is the next step.

Multimeters and Voltmeters

Voltage measurements are taken to verify that expected voltage levels are present at key points in the circuit. Measure the voltage at different nodes using a multimeter set to the appropriate voltage range. Compare these readings with the expected values based on the circuit design. Discrepancies can reveal issues such as power supply problems or faulty components.

Because of that need to measure, multimeters are indispensable tools for measuring electrical properties like voltage, current, and resistance. They come in two main varieties: analog and digital.

When is an analog multimeter preferred to a digital multimeter? (G4B09)

A. When testing logic circuits
B. When high precision is desired
C. When measuring the frequency of an oscillator
D. When adjusting circuits for maximum or minimum values

An analog multimeter is preferred when adjusting circuits for maximum or minimum values because it provides a continuously moving needle. This visual feedback helps users make fine adjustments more smoothly than digital readouts, which can fluctuate or update less fluidly.

What is an advantage of a digital multimeter as compared to an analog multimeter? (G4B06)

A. Better for measuring computer circuits
B. Less prone to overload
C. Higher precision
D. Faster response

However, digital multimeters have their advantages, especially when precision is required. They offer higher accuracy and are better suited for exact measurements, providing clear numerical values without the interpretation needed for an analog display.

Voltmeters have high input impedance, which is another benefit. This reduces the loading effect on the circuit being measured, minimizing any impact the meter itself might have on the circuit's behavior. This feature makes them ideal for sensitive electronic components.

Why do voltmeters have high input impedance? (G4B05)

A. It improves the frequency response
B. It allows for higher voltages to be safely measured
C. It improves the resolution of the readings
D. It decreases the loading on circuits being measured

Multimeters—both analog and digital—each has its place. Analog versions shine in tasks where adjustments for maximum or minimum values are needed, while digital multimeters, due to their high input impedance, provide greater precision and reduce circuit loading. Understanding when to use each type is key to effective circuit troubleshooting and optimization.

Oscilloscope

An oscilloscope is invaluable for visualizing waveforms and detecting issues that are not apparent through simple voltage measurements. An oscilloscope lets you see distortions, noise, and other anomalies by displaying the signal in real time. A signal generator can be used to inject test signals into the circuit, helping you trace and diagnose problems. Keeping a log of measurements and observations is also beneficial. Documenting your findings provides a reference for future troubleshooting and helps track changes over time.

The oscilloscope contains horizontal and vertical channel amplifiers, allowing it to display and measure electrical signals in real-time. The vertical channel represents the signal's voltage, while the horizontal channel represents time. By plotting the voltage against time, you can visualize waveforms and understand how a circuit is behaving. This makes the oscilloscope incredibly useful for diagnosing problems, monitoring signals, or observing complex waveforms that simpler tools like digital voltmeters cannot capture.

What item of test equipment contains horizontal and vertical channel amplifiers? (G4B01)

A. An ohmmeter
B. A signal generator
C. An ammeter
D. An oscilloscope

In addition to displaying basic voltage signals, oscilloscopes are particularly helpful in examining more complex waveforms, such as those generated by a CW (continuous wave) transmitter. When checking the keying waveform of a CW transmitter, an oscilloscope is the best tool for the job because it allows you to see the shape and timing of the keying pulses, ensuring that the transmitter is functioning correctly. This is especially important because distorted keying can lead to communication errors or signal interference.

Which of the following is the best instrument to use for checking the keying waveform of a CW transmitter? (G4B03)

A. An oscilloscope
B. A field strength meter
C. A sidetone monitor
D. A wavemeter

Another advantage of using an oscilloscope compared to a digital voltmeter is that it can display complex waveforms that a voltmeter could not accurately measure. While a voltmeter can only give you a numerical readout of voltage, the oscilloscope provides a visual representation, making it a superior tool for observing RF signals, such as when checking the RF envelope pattern of a transmitted signal by connecting the attenuated RF output of the transmitter to the vertical input.

Which of the following is an advantage of an oscilloscope versus a digital voltmeter? (G4B02)

A. An oscilloscope uses less power
B. Complex impedances can be easily measured
C. Greater precision
D. Complex waveforms can be measured

What signal source is connected to the vertical input of an oscilloscope when checking the RF envelope pattern of a transmitted signal? (G4B04)

A. The local oscillator of the transmitter
B. An external RF oscillator
C. The transmitter balanced mixer output
D. The attenuated RF output of the transmitter

Oscilloscopes are tools for visualizing electrical signals, particularly when measuring complex waveforms or monitoring the behavior of RF signals. They offer capabilities beyond simple voltage measurements, making them ideal for tasks like checking the keying waveform of a CW transmitter or examining RF envelope patterns.

Chapter Summary

Mastering circuit diagrams is an essential skill for any radio enthusiast, providing a clear roadmap for understanding, building, and troubleshooting the circuits that power a radio station. Recognizing schematic symbols and using diagnostic tools like multimeters, voltmeters, and oscilloscopes equips you to confidently address common issues, deepen your understanding of electronic circuits, and enhance your practical skills. By systematically tackling circuit challenges and employing effective troubleshooting techniques, you ensure the reliable operation of your ham radio equipment while strengthening your technical proficiency.

This solid foundation keeps your station running smoothly and prepares you for the challenges and opportunities that come with advancing your license.

Chapter 12

Mobile Stations

OPERATING A HAM RADIO station from a mobile setup offers unique opportunities and challenges. Mobile stations allow amateur radio operators to communicate on the go, such as from a car, boat, or other vehicles. This mobility is ideal for participating in emergency communications, DXing, or simply enjoying the flexibility of radio communication during travel. However, mobile setups often require a balance between convenience and performance.

One of the biggest challenges in mobile operations is the antenna. Unlike home-based stations, where operators can use full-sized antennas, mobile stations must use compact or shortened antennas due to vehicle space constraints. These shortened antennas are less efficient and provide limited bandwidth, which can reduce signal strength and require frequent tuning. Additionally, mobile stations often face interference from the vehicle's electrical systems, such as the battery charging system and onboard computers. Understanding and addressing these challenges is critical to maintaining effective communication while on the move.

Mobile Antennas

One of the most significant challenges in an HF mobile installation is the antenna's efficiency, mainly because mobile antennas tend to be electrically short. An electrically short antenna is one that is physically shorter than the optimal length needed for the frequency being used, which reduces its efficiency. The reduced efficiency happens because shorter antennas are less able to radiate energy effectively, resulting in weaker signal transmission and reception. In HF mobile setups, this limitation is critical since vehicles often require compact antennas. Therefore, the antenna's efficiency directly impacts the performance of the mobile station.

> **Which of the following most limits an HF mobile installation? (G4E05)**
>
> A. "Picket fencing"
> B. The wire gauge of the DC power line to the transceiver
> **C. Efficiency of the electrically short antenna**
> D. FCC rules limiting mobile output power on the 75-meter band

While convenient for vehicle installation, a shortened mobile antenna has a notable trade-off: a very limited operating bandwidth. This means the antenna is efficient only over a narrow range of frequencies, and tuning becomes necessary when switching between different parts of the HF spectrum. Full-sized antennas, by contrast, offer a much broader bandwidth, allowing them to cover a wider range of frequencies without requiring constant adjustments. For mobile users, this limitation can be frustrating, as it reduces the flexibility of the station's operation.

What is one disadvantage of using a shortened mobile antenna as opposed to a full-size antenna? (G4E06)

A. Short antennas are more likely to cause distortion of transmitted signals
B. Q of the antenna will be very low
C. Operating bandwidth may be very limited
D. Harmonic radiation may increase

Both of these questions highlight the inherent challenges of mobile HF operation, particularly focusing on antenna size and efficiency. Mobile antennas are often compromised in size, affecting their efficiency and bandwidth, which mobile operators must consider when setting up their stations.

Mobile Antenna Components

Mobile antennas have several components designed to optimize the performance of shortened antennas.

Capacitance Hat

A capacitance hat is a device used on a mobile antenna to "electrically lengthen" a physically short antenna. Since a full-sized HF antenna is often too long for practical vehicle use, a shortened one is used instead. The capacitance hat compensates for the lack of physical length by increasing the antenna's electrical length, making it behave more like a full-sized one. This improves the efficiency of the antenna, allowing it to radiate more power effectively. In simpler terms, it tricks the antenna into thinking it's longer than it really is, which helps it perform better despite its compact size.

What is the purpose of a capacitance hat on a mobile antenna? (G4E01)

A. To increase the power handling capacity of a whip antenna
B. To reduce radiation resistance
C. To electrically lengthen a physically short antenna
D. To lower the radiation angle

Hat

Corona Ball

The corona ball is another important component found on mobile antennas, particularly on HF antennas. It is a small, rounded metallic ball placed at the tip of the antenna. <u>Its purpose is to reduce RF voltage discharge, or "corona discharge," from the sharp tip of the antenna while transmitting.</u> Without the corona ball, high RF voltages can build up at the tip, which can lead to interference, energy loss, or even cause sparks in some cases. The corona ball smooths out this voltage, reducing the chance of unwanted discharge and improving transmissions.

Ball

What is the purpose of a corona ball on an HF mobile antenna? (G4E02)

A. To narrow the operating bandwidth of the antenna
B. To increase the "Q" of the antenna
C. To reduce the chance of damage if the antenna should strike an object
D. To reduce RF voltage discharge from the tip of the antenna while transmitting

The capacitance hat and corona ball help improve the functionality of mobile antennae despite their size and the challenges of operating in a mobile environment.

Power Connections in HF Mobile Installations

Power connections in HF mobile installations are critical for ensuring reliable, efficient, and safe operation. A 100-watt HF mobile transceiver draws significant current, often exceeding 20 amps during peak transmission. This power demand level means your power connection must be designed to handle high current without causing issues like voltage drops, overheating, or interference.

To achieve this, appropriately sized wires, typically at least 10-gauge or thicker, is needed to minimize resistance and avoid excessive heat buildup. Always connect directly to the vehicle's battery, as this provides the most stable and clean power source. Avoid using the vehicle's accessory power outlets, which often cannot handle the current required by an HF transceiver and may introduce electrical noise.

Properly fuse positive and negative leads as close to the battery as possible to protect against short circuits. Ensure all connections are secure, and use high-quality terminals and connectors to reduce the risk of power loss or sparking. Routing power cables away from other electrical components can minimize interference and maintain signal clarity.

By following these practices, you can ensure safe and reliable power delivery to your HF mobile radio, maximizing performance and minimizing potential issues during operation.

Direct, Fused Power Connection

<u>The best method for powering a 100-watt HF mobile transceiver is to use a direct connection to the vehicle's battery with a heavy-gauge wire protected by a fuse.</u> This approach ensures that the transceiver receives a stable supply of power without

interference from other vehicle systems. The fuse is critical for protecting the equipment and vehicle in case of a short circuit or power surge. Heavy-gauge wire is necessary to handle the high current drawn by the transceiver during transmission, ensuring that the radio operates efficiently without causing a voltage drop, which could affect performance.

Which of the following direct, fused power connections would be the best for a 100-watt HF mobile installation? (G4E03)

A. To the battery using heavy-gauge wire
B. To the alternator or generator using heavy-gauge wire
C. To the battery using insulated heavy duty balanced transmission line
D. To the alternator or generator using insulated heavy duty balanced transmission line

---

Avoiding the Vehicle's Auxiliary Power Socket

Many mobile operators might be tempted to use the vehicle's auxiliary power socket (commonly the cigarette lighter socket) to supply power to their transceiver. However, this is not recommended for a 100-watt HF transceiver. The wiring in the auxiliary socket is typically not designed to handle the high current demands of the transceiver, leading to inadequate power delivery, potential overheating, or blown fuses. Instead, a dedicated power connection to the battery ensures the transceiver gets the power it needs for effective operation without compromising safety or performance.

Why should DC power for a 100-watt HF transceiver not be supplied by a vehicle's auxiliary power socket? (G4E04)

A. The socket is not wired with an RF-shielded power cable
B. The socket's wiring may be inadequate for the current drawn by the transceiver
C. The DC polarity of the socket is reversed from the polarity of modern HF transceivers
D. Drawing more than 50 watts from this socket could cause the engine to overheat

---

The direct power connection using the correct wire gauge and fuse ensures that your mobile HF radio setup will be both safe and reliable in your mobile operations.

Mobile Interference

Mobile installations present a unique set of interference challenges. Vehicles' internal electronics can severely degrade HF reception if proper precautions aren't taken.

Which of the following may cause receive interference to an HF transceiver installed in a vehicle? (G4E07)

A. The battery charging system
B. The fuel delivery system
C. The control computers
D. All these choices are correct

When installing a HF transceiver in a vehicle, various systems within the car can cause interference, affecting your ability to receive signals. The most common sources of interference include the vehicle's battery charging system, fuel delivery system, and control computers. Each of these systems generates electrical noise that can be picked up by the HF transceiver, leading to distorted or unclear signals. For example, alternators and fuel injectors can create pulsing noises, while onboard computers can generate broad-spectrum noise that interferes with your radio. Understanding and addressing these potential sources of interference is key to ensuring a clean and reliable mobile HF setup.

Chapter Summary

This chapter covered the exam questions on setting up and optimizing mobile and portable radio stations. From understanding the components of mobile antennas to learning about secure power connections and managing interference, each aspect is fundamental for achieving reliable, clear communication on the go. With this knowledge, you can be assured of strong signals and reliable operations wherever your mobile radio takes you.

Chapter 13

Feedlines

IN HAM RADIO AND other radio communications, feedlines are the cables or wires that carry radio frequency (RF) energy from the transmitter to the antenna and back to the receiver. They ensure efficient power transfer with minimal signal loss or interference.

Impedance Matching

Impedance matching ensures that the maximum amount of power is transferred from the transmitter to the antenna without unnecessary loss. This section will explore how impedance matching improves communication efficiency and clarity.

What is one reason to use an impedance matching transformer at a transmitter output? (G7C03)

A. To minimize transmitter power output
B. To present the desired impedance to the transmitter and feed line
C. To reduce power supply ripple
D. To minimize radiation resistance

An impedance matching transformer is used at a transmitter's output to match the transmitter's impedance to the feed line or antenna. This ensures efficient power transfer, minimizing signal loss due to reflection or mismatch. Without proper impedance matching, a portion of the signal would be reflected back to the transmitter, causing inefficiencies and potentially damaging equipment. The transformer helps balance the system for optimal performance.

Devices like a transformer, a Pi-network, and a length of transmission line can be used to adjust or match the impedance between components at radio frequencies, improving signal efficiency.

- **Transformers** work by adjusting the voltage and current ratios, allowing you to match different impedance levels between circuit stages.

- **Pi-networks** use a combination of capacitors and an inductor to tune the circuit and match impedances, which are commonly used in amplifiers or antenna tuners.

- **A length of transmission line** can also be used to match impedance by adjusting its length to a specific wavelength, which ensures that signals travel efficiently without reflections.

Which of the following devices can be used for impedance matching at radio frequencies? (G5A10)

A. A transformer
B. A Pi-network
C. A length of transmission line
D. All these choices are correct

To remember this for the exam, consider these devices as "adjusters" that help optimize the connection between parts of a radio circuit, ensuring the signal flows smoothly and efficiently without losses due to mismatched impedance.

Feedlines

Feedlines come in different types, each with varying characteristics of impedance, attenuation (signal loss), and frequency capabilities. The most common types are:

1. **Coaxial Cable (Coax)**: The most widely used feedline, consisting of a central conductor surrounded by insulation, a shield (usually braided wire or foil), and an outer insulating jacket. It's easy to install and provides good shielding from external interference. Coaxial cables typically have impedances of 50 ohms or 75 ohms.

2. **Parallel Conductor Feedlines (Ladder Line or Window Line)**: This type consists of two parallel wires separated by spacers. These lines have much lower loss than coax, particularly at higher frequencies, but are more susceptible to environmental factors like moisture. Typical impedances are 300 ohms and 450 ohms.

The choice of feedline depends on factors like the operating frequency, the distance between the radio and the antenna, and the need for impedance matching to minimize the standing wave ratio (SWR), which affects signal transmission efficiency.

Feedlines bridge your transmitter/receiver and the antenna, ensuring the signal is carried with minimal loss. The right type of feedline ensures efficient operation by matching impedance and reducing losses.

Parallel Conductor Feedline

A parallel conductor feedline is a type of transmission line used to carry RF energy from a transmitter to an antenna. It consists of two parallel conductors, spaced apart by insulating material, which allows it to transfer signals efficiently over long distances.

Parallel conductor feedlines are ideal for efficient, multi-band antenna systems or situations where minimizing loss is crucial. They're a great choice for balanced antennas like dipoles, especially when paired with a tuner to handle impedance mismatches.

The characteristic impedance of a parallel conductor feed line is determined by two key factors: the distance between the centers of the two conductors and the radius of each conductor. Simply put, the farther apart the conductors are, the higher the impedance will be—likewise, the larger the conductor radius, the lower the impedance. The characteristic impedance is important because it helps determine how efficiently power is transferred from the transmitter to the antenna without creating reflections or losses.

Which of the following factors determine the characteristic impedance of a parallel conductor feed line? (G9A01)

A. The distance between the centers of the conductors and the radius of the conductors
B. The distance between the centers of the conductors and the length of the line
C. The radius of the conductors and the frequency of the signal
D. The frequency of the signal and the length of the line

For a parallel conductor feed line, the spacing between the conductors and the size of the conductors play a critical role in determining its characteristic impedance. Ensuring proper impedance matching can minimize transmission losses and maximize power transfer.

Window Line

Window line is a type of parallel conductor feedline used to carry RF signals between a transmitter and an antenna. It gets its name from the small "windows" or cutouts in the insulating material that separates the two conductors. These cutouts reduce weight and material while maintaining the proper spacing between the wires.

Window line is an excellent choice for operators seeking efficient feedline solutions, especially in high-frequency or multi-band setups where low loss and versatility are priorities. It's often paired with balanced antennas and antenna tuners to maximize performance. Its nominal characteristic impedance is 450 ohms.

What is the nominal characteristic impedance of "window line" transmission line? (G9A03)

A. 50 ohms
B. 75 ohms
C. 100 ohms
D. 450 ohms

The window line's 450-ohm nominal characteristic impedance makes it a popular feedline choice for HF communications. The open windows in the line help reduce losses while maintaining a manageable impedance level.

Attenuation and Frequency Effects

When transmitting signals from your radio to an antenna, it's essential to understand how energy losses occur in the feedline. Attenuation refers to the reduction of signal strength as it travels along the feedline, and this is influenced by factors such as the frequency of the signal and the type of feedline used. As the frequency increases, attenuation (or signal loss) generally increases, making it crucial to choose appropriate feedline types and lengths to minimize these losses. This section covers the effects of attenuation and frequency on feedlines and how RF feedline loss is measured and expressed.

How does the attenuation of coaxial cable change with increasing frequency? (G9A05)

A. Attenuation is independent of frequency
B. Attenuation increases
C. Attenuation decreases
D. Attenuation follows Marconi's Law of Attenuation

The attenuation of a coaxial cable refers to the amount of signal loss that occurs as the radio frequency signal travels through the cable. This loss becomes more pronounced as the frequency of the signal increases. This is because higher frequencies result in more energy being absorbed by the dielectric material inside the cable and more power being lost as heat. In practical terms, this means that if you're operating at higher frequencies, you may experience more significant signal degradation over long distances unless you use a low-loss cable specifically designed to minimize this effect.

Higher frequency signals experience greater attenuation, meaning more signal strength is lost over a given distance. This is an key factor to consider when designing a communication system, especially for high-frequency operations.

In what units is RF feed line loss usually expressed? (G9A06)

A. Ohms per 1,000 feet
B. Decibels per 1,000 feet
C. Ohms per 100 feet
D. Decibels per 100 feet

For practical purposes, RF feed line loss is typically measured in decibels (dB) over a given distance, commonly per 100 feet. The decibel is a logarithmic unit that quantifies the ratio of the power loss between two points along the feedline. A loss of 1 dB means that approximately 20% of the signal's power is lost, while a loss of 3 dB represents a 50% power loss. This measurement helps operators compare different types of feedlines and determine how much signal will be lost over a specific length of cable at a given frequency. The lower the dB value per 100 feet, the better the cable is at preserving signal strength over that distance.

Feed line loss is measured in decibels per 100 feet, which allows you to estimate the signal you will lose over a specific cable length.

Attenuation and signal loss are important considerations when planning your radio system. As frequency increases, attenuation becomes more significant, so choosing the right feedline with minimal loss is crucial for maintaining signal strength. Understanding how to measure loss in decibels per 100 feet gives you a practical way to select and optimize your feedline setup. In doing so, you can ensure that your signal reaches its intended destination with as little degradation as possible.

Chapter Summary

Feedlines are the critical link between your radio and antenna, ensuring efficient power transfer with minimal signal loss. Understanding the characteristics of different feedlines, such as coaxial cables and parallel conductor feedlines like window line, allows you to select the best option for your operating needs.

Key factors such as impedance, attenuation, and frequency effects influence feedline performance, making it important to choose and manage feedlines carefully to maintain signal strength. By recognizing how attenuation increases with frequency and measuring feedline loss in decibels per 100 feet, you can optimize your station for reliable and efficient communication. With the right feedline setup, your signals will travel farther and clearer, enhancing your overall operating experience.

Chapter 14

Antennas

YOUR ANTENNA IS THE gateway to global communication. Whether you're reaching out to local operators or making long-distance contacts, your antenna plays a pivotal role in your ability to transmit and receive signals. As you upgrade to General Class, having a deeper understanding of the different types of antennas and their specific characteristics becomes more important than it was with just a Technician's license.

Random Wire

A random wire HF antenna is a simple, versatile antenna made from a single length of wire, typically connected to an antenna tuner and used for transmitting and receiving on HF bands. It is called "random" because its length is not cut to a specific fraction of a wavelength for a particular frequency, making it usable across multiple bands with the help of a tuner.

What is a characteristic of a random-wire HF antenna connected directly to the transmitter? (G9B01)

A. It must be longer than 1 wavelength
B. Station equipment may carry significant RF current
C. It produces only vertically polarized radiation
D. It is more effective on the lower HF bands than on the higher bands

A random-wire antenna is an unbalanced, non-resonant antenna often used when space or other constraints limit antenna options. When connected directly to the transmitter without any matching or balancing network, significant RF currents may flow through the station's equipment, leading to potential RF interference or safety issues. This is due to the mismatched impedance between the antenna and the transmitter, causing RF currents to flow where they shouldn't, such as along the feed line or chassis of the equipment.

A random wire HF antenna is a versatile and cost-effective solution for multi-band HF operation, especially when simplicity and portability are key. It's easy to set up and ideal for operators with limited space or those operating in the field. While an antenna tuner and a good RF ground are essential for optimal performance, this antenna offers excellent flexibility across multiple bands. To ensure safe and efficient operation, it's crucial to properly match the impedance and establish a reliable ground to minimize RF currents in station equipment.

Monopole

A monopole antenna is a type of radio antenna consisting of a single conductive element—typically a rod or wire. It is called a "monopole" because it has only one pole or element, unlike a dipole antenna, which has two elements. Monopole antennas are popular due to their simplicity, compact size, and effectiveness in applications where vertical polarization and omnidirectional coverage are beneficial.

Quarter-wave

Quarter-waves are the most popular monopoles due to their balance of efficiency and simplicity. These antennas play a significant role in both amateur and professional radio setups.

Which of the following is a common way to adjust the feed point impedance of an elevated quarter-wave ground-plane vertical antenna to be approximately 50 ohms? (G9B02)

A. Slope the radials upward
B. Slope the radials downward
C. Lengthen the radials beyond one wavelength
D. Coil the radials

In a quarter-wave ground-plane vertical antenna, adjusting the angle of the radials (wires extending from the base of the antenna) can help match the antenna's feed point impedance to that of a standard 50-ohm coaxial cable. Slope the radials downward at an angle of around 45 degrees to lower the impedance to approximately 50 ohms, which is ideal for ensuring maximum power transfer between the transmitter and the antenna.

Adjusting the radials downward is a practical way to improve the impedance match of a quarter-wave vertical antenna, optimizing the antenna's efficiency and reducing signal loss.

Which of the following best describes the radiation pattern of a quarter-wave ground-plane vertical antenna? (G9B03)

A. Bi-directional in azimuth
B. Isotropic
C. Hemispherical
D. Omnidirectional in azimuth

A quarter-wave ground-plane vertical antenna radiates equally in all directions (omnidirectional in azimuth) along the horizontal plane, meaning the signal strength is uniform regardless of direction. This makes it ideal for applications like local communication, where signals need to reach receivers in all directions without favoring one particular angle over another.

The omnidirectional radiation pattern makes the quarter-wave vertical antenna versatile for various communication needs, particularly for covering a wide area equally.

Ok, I know what azimuth means. But can you explain it to me just one more time?

In the context of radio communication and antennas, azimuth refers to the horizontal angle or direction of an antenna's radiation pattern relative to a reference point, typically true north. It is measured in degrees, with 0° representing north, 90° east, 180° south, and 270° west. The azimuth angle helps to determine where the strongest signal is transmitted or received along the horizontal plane.

For example, in an omnidirectional antenna, the azimuth pattern would be a 360° circle, indicating equal radiation in all directions horizontally. In directional antennas, like a Yagi, the azimuth pattern might show a stronger signal in one specific direction, which is helpful for long-distance communications where focusing the signal is important.

In summary, azimuth helps operators understand how an antenna radiates or receives signals across the horizontal plane, aiding in positioning for optimal communication.

What is the approximate length for a 1/4 wave monopole antenna cut for 28.5 MHz? (G9B12)

A. 8 feet
B. 11 feet
C. 16 feet
D. 21 feet

Step 1: Use the Formula for Quarter-Wave Antenna Length

The formula to calculate the length of a quarter-wave antenna is:

$$Antenna\ Length\ (feet) = \frac{468\ feet}{Frequency\ (in\ MHz)}$$

Where:

- L = Length of the antenna in feet

- f = Frequency in megahertz (MHz)

Step 2: Plug in the Frequency & Solve

The given frequency is 28.5 MHz. Substituting into the formula:

$$Antenna\ Length = \frac{468}{28.5} \approx 16.4\ feet$$

$$16.4\ feet\ divided\ by\ two\ to\ get\ quarter\ wave\ = 8.2\ feet$$

L = 8.2 feet rounded equals 8 feet!

This calculation ensures that the antenna is the correct length for efficient resonance and transmission at that frequency. Knowing how to calculate the appropriate antenna length based on frequency is crucial for ensuring effective communication and minimizing signal loss.

Understanding how to adjust elements such as radial angles and length to match the operating frequency allows amateur radio operators to maximize their antenna's performance. Whether setting up for local or long-range communication, these foundational concepts in antenna design ensure reliable and clear signals.

Vertical

Vertical antennas are versatile and widely used, known for their compact design and effective performance.

Vertical antennas are often implemented as quarter-wave monopoles, combining the simplicity of the monopole design with vertical polarization for efficient omnidirectional coverage, making them ideal for local coverage and ground-wave propagation while also performing well for long-distance (DX) communication on lower HF bands.

Their vertically polarized signals minimize ground losses, and their straightforward design makes them suitable for setups with limited space or portable operations. Whether used in base stations or temporary field setups, vertical antennas offer a practical balance of simplicity, efficiency, and effectiveness for ham radio enthusiasts.

Where should the radial wires of a ground-mounted vertical antenna system be placed? (G9B06)

A. As high as possible above the ground
B. Parallel to the antenna element
C. On the surface or buried a few inches below the ground
D. At the center of the antenna

Radial wires in a ground-mounted vertical antenna system act as part of the ground system, helping to improve signal efficiency by creating a low-resistance path for the return current. These wires should be placed directly on the ground's surface or buried just a few inches below it. Burying them deeper isn't necessary and can make installation harder without providing added benefit. A well-laid radial system helps minimize ground losses, improving antenna performance, particularly for long-distance (DX) communication.

Which of the following is an advantage of using a horizontally polarized as compared to a vertically polarized HF antenna? (G9B09)

A. Lower ground losses
B. Lower feed point impedance
C. Shorter radials
D. Lower radiation resistance

Horizontal polarization typically results in lower ground losses compared to vertical polarization. This is because horizontally polarized waves interact with the ground in a way that reduces power absorption, allowing more energy to be radiated into the air. On the other hand, vertically polarized antennas tend to have higher ground losses since the signal must travel through the ground before radiating into the atmosphere. For HF communications, lower ground losses can significantly improve efficiency, making horizontally polarized antennas a better choice in many situations.

Dipole

Dipole antennas are among ham radio's most fundamental and widely used antennas. These antennas consist of two conductive elements, each half the length of the desired wavelength, making them versatile for various HF bands. The dipole antenna's simplicity makes it an excellent choice for beginners and experienced operators. It is straightforward to construct and requires basic materials like wire, insulators, and a center feed point. When installed at an appropriate height, typically half a wavelength above the ground, dipole antennas offer excellent local and medium-range communication performance. However, they require sufficient space for installation, which can be a limitation in urban or restricted environments.

Basic Characteristics of Dipole

What is the radiation pattern of a dipole antenna in free space in a plane containing the conductor? (G9B04)

A. It is a figure-eight at right angles to the antenna
B. It is a figure-eight off both ends of the antenna
C. It is a circle (equal radiation in all directions)
D. It has a pair of lobes on one side of the antenna and a single lobe on the other side

A dipole antenna exhibits a figure-eight radiation pattern when placed in free space. This means that most of the radiation is concentrated broadside to the antenna, while little to no radiation occurs off the ends. The pattern resembles a figure-eight, where the strongest signals are emitted at right angles to the antenna, making it ideal for long-range communication in those directions.

A dipole antenna's radiation pattern is highly directional, with maximum radiation occurring perpendicular to the antenna's wire. This figure-eight pattern is essential for focusing signals and optimizing communication range in the desired directions.

Due to their simplicity and efficiency, dipole and monopole antennas are widely used. Understanding how these antennas behave in terms of RF currents, radiation patterns, and impedance matching helps ensure safe and effective operation. Knowing the basics allows operators to make informed decisions about their antenna setup and optimize communication performance, whether using a random-wire HF antenna or a precisely tuned dipole.

Azimuthal Radiation Pattern

Dipole antennas are one of the most fundamental and widely used antennas in amateur radio due to their simplicity and efficiency. Understanding their characteristics, such as radiation patterns and impedance, helps in optimizing their performance for various frequencies and environments. The dipole's behavior, including how its height above ground and feed point location affect its impedance and radiation pattern, plays a crucial role in ensuring effective transmission and reception. These considerations are important for setting up reliable HF communications, especially for General Class operators.

How does antenna height affect the azimuthal radiation pattern of a horizontal dipole HF antenna at elevation angles higher than about 45 degrees? (G9B05)

A. If the antenna is too high, the pattern becomes unpredictable
B. Antenna height has no effect on the pattern
C. If the antenna is less than 1/2 wavelength high, the azimuthal pattern is almost omnidirectional
D. If the antenna is less than 1/2 wavelength high, radiation off the ends of the wire is eliminated

When a horizontal dipole antenna is placed closer to the ground (less than 1/2 wavelength high), its radiation pattern becomes nearly omnidirectional in the horizontal plane. This means it radiates signals in all directions more evenly. At higher elevation angles, the antenna sends energy more upward than outward. This is useful for short-range or "local" contacts where higher elevation angles are advantageous. For long-distance communication (DX), installing the dipole at a higher elevation is typically better for producing lower-angle radiation, favoring far-reaching signals.

Feed Point Impedance

Feed point impedance is the electrical resistance and reactance measured at the point where the transmission line connects to an antenna. It represents how much the antenna resists the flow of RF current from the feedline and is typically expressed in ohms. By understanding and matching the impedance, operators can minimize power losses, reduce SWR, and optimize antenna performance for better communication.

How does the feed point impedance of a horizontal 1/2 wave dipole antenna change as the antenna height is reduced to 1/10 wavelength above ground? (G9B07)

A. It steadily increases
B. It steadily decreases
C. It peaks at about 1/8 wavelength above ground
D. It is unaffected by the height above ground

As the height of a 1/2 wave dipole decreases relative to the ground (down to 1/10 wavelength), the impedance at the feed point drops significantly. A dipole placed higher above ground tends to have an impedance closer to the theoretical 73 ohms, but lowering it causes the antenna to interact more with the ground, reducing its impedance. This lower impedance can sometimes necessitate matching networks to prevent high SWR and improve the overall efficiency of the antenna.

How does the feed point impedance of a 1/2 wave dipole change as the feed point is moved from the center toward the ends? (G9B08)

A. It steadily increases
B. It steadily decreases
C. It peaks at about 1/8 wavelength from the end
D. It is unaffected by the location of the feed point

The feed point impedance of a 1/2 wave dipole is typically around 73 ohms when fed at the center. However, as you move the feed point towards the ends of the antenna, the impedance increases, reaching much higher values near the ends. This is because the current distribution along the dipole is highest in the center and drops towards the ends while the voltage increases. End-feeding a dipole results in very high impedance, often requiring impedance matching to avoid significant SWR issues.

An end-fed half-wave (EFHW) antenna is exactly what it sounds like—a half-wavelength long antenna fed at one of its ends instead of the center. Unlike a center-fed dipole, where the impedance at the center is around 50 to 75 ohms, the feed point impedance of an EFHW antenna is very high, typically in the range of several thousand ohms. This happens because the voltage at the end of a half-wave antenna is at a maximum while the current is very low, creating a high-impedance point.

What is the feed point impedance of an end-fed half-wave antenna? (G9D02)

A. Very low
B. Approximately 50 ohms
C. Approximately 300 ohms
D. **Very high**

To use an EFHW antenna effectively an impedance matching transformer is used to bring the impedance down to a more manageable level. Without proper matching, a radio connected directly to an EFHW would see excessive standing wave ratio (SWR), making the system inefficient and potentially damaging the transmitter. Remembering this concept will help you recognize why EFHW antennas require matching components and why their feed point impedance is so high.

Dipole Length

What is the approximate length for a 1/2 wave dipole antenna cut for 14.250 MHz? (G9B10)

A. 8 feet
B. 16 feet
C. 24 feet
D. 33 feet

To calculate the length of a 1/2 wave dipole, you can use the formula:

$$Length\ (in\ feet) = \frac{468\ feet}{Frequency\ (in\ MHz)}$$

Does this formula look familiar? It should be from the Technician's. It's also twice the length of what we just saw in the quarter-wave section, which makes sense going from quarter to half-wave!

Substitute into the formula and solve:

$$Length\ (in\ feet) = \frac{468\ feet}{14.250\ MHz} \approx 32.8\ or\ round\ to\ 33\ feet$$

The length of 14.250 MHz is approximately 33 feet. This formula helps determine the optimal size of the antenna to achieve resonance on the desired frequency band, ensuring efficient transmission and reception.

Try another.

> **What is the approximate length for a 1/2 wave dipole antenna cut for 3.550 MHz? (G9B11)**
>
> A. 42 feet
> B. 84 feet
> **C. 132 feet**
> D. 263 feet

Using the same dipole formula:

$$Length\ (in\ feet) = \frac{468\ feet}{Frequency\ (in\ MHz)}$$

Substitute into the formula:

$$Length\ (in\ feet) = \frac{468\ feet}{3.550\ MHz} \approx 131.8\ or\ round\ to\ 132\ feet$$

This calculation gives a dipole length of around 132 feet. The required antenna length is much longer at lower frequencies, like 3.550 MHz (commonly in the 80-meter band). Operators may need to adjust the installation to accommodate the space required for such large antennas or use techniques like loading coils to shorten the physical length while maintaining electrical resonance.

Dipole antennas, whether used at high or low frequencies, present unique challenges regarding installation height, impedance matching, and overall effectiveness. Understanding the relationships between height, feed point location, and impedance is crucial for optimizing antenna performance. By knowing the correct calculations and characteristics, operators can ensure their antennas are tuned correctly for effective communication across various bands.

Inverted V

The inverted V antenna is a variation of the traditional dipole antenna with a single central support. Instead of being mounted horizontally, the antenna elements slope downwards from a central support, forming a "V" shape. This configuration is advantageous when space is limited, as it requires less horizontal area. The inverted V also offers a lower radiation angle than a standard dipole, which can enhance long-distance communication capabilities.

What is the common name of a dipole with a single central support? (G9D12)

A. Inverted V
B. Inverted L
C. Sloper
D. Lazy H

The inverted V dipole is a versatile and space-efficient antenna option for amateur radio operators, balancing ease of installation with solid all-around performance. Its flexibility makes it a popular choice for both fixed and portable operations.

Setting Up and Optimizing Your Antenna

Antenna Tuner

An antenna tuner is a device used in radio communications to improve the efficiency of power transfer from the transmitter to the antenna system through the feed line. In practical terms, the antenna tuner adjusts the impedance seen by the transmitter to match the feed line and antenna impedance. This is crucial because if the impedance of the antenna system is not correctly matched to the transmitter's output, a significant amount of power could be reflected back to the transmitter, causing poor performance and potential damage.

What is the purpose of an antenna tuner? (G4A06)

A. Reduce the SWR in the feed line to the antenna
B. Reduce the power dissipation in the feedline to the antenna
C. Increase power transfer from the transmitter to the feed line
D. All these choices are correct

Using an antenna tuner maximizes the amount of power radiated by the antenna rather than being reflected. This is particularly helpful when using antennas that operate on multiple frequencies, as the tuner helps match the impedance even when the antenna is not perfectly resonant at the desired operating frequency. Essentially, the tuner acts as a mediator between the transmitter and the antenna, ensuring efficient power transfer and better overall performance.

Antenna Analyzer

An antenna analyzer is a versatile tool amateur radio operators use to measure various parameters of antennas and transmission lines. One of the features of an antenna analyzer is its ability to measure the impedance of coaxial cables. Impedance is critical in ensuring efficient power transfer between the transmitter and antenna. By checking the impedance, an operator can verify that the feed line (coaxial cable) matches the antenna and transmitter correctly, reducing losses and preventing signal reflections. If there is a mismatch in impedance, power will be reflected back toward the transmitter, potentially causing inefficiencies and damage.

Which of the following can be measured with an antenna analyzer? (G4B13)

A. Front-to-back ratio of an antenna
B. Power output from a transmitter
C. Impedance of coaxial cable
D. Gain of a directional antenna

Impedance measurements with an antenna analyzer are necessary because they allow operators to fine-tune their transmission system for maximum performance. Coaxial cable impedance is usually specified (e.g., 50 ohms), and ensuring that this matches the antenna and transmitter prevents unnecessary power loss. Using an analyzer to measure impedance can also help diagnose issues like cable degradation, poor connections, or faulty components in the feed line.

An antenna analyzer is crucial for measuring the impedance of coaxial cables and ensuring that the entire transmission system is working efficiently. By matching the impedance of the cable, antenna, and transmitter, operators can optimize performance, reduce power loss, and avoid damaging their equipment.

Standing Wave Ratio (SWR)

In radio communication, SWR (Standing Wave Ratio) is a critical factor that determines how efficiently power is transferred from the transmitter to the antenna. SWR is affected by impedance mismatches between the feed line and the antenna. When this mismatch occurs, standing waves form on the feed line, causing signal reflections, which reduce the effectiveness of signal transmission. This section explores the causes and solutions for standing waves and discusses how transmission line losses can impact SWR measurements.

Which of the following can be determined with a directional wattmeter? (G4B10)

A. Standing wave ratio
B. Antenna front-to-back ratio
C. RF interference
D. Radio wave propagation

A directional wattmeter is a tool amateur radio operators use to measure the standing wave ratio (SWR). The SWR is a critical parameter that indicates how efficiently power is being transferred from the transmitter to the antenna through the feed line. A low SWR means that most of the power is reaching the antenna, while a high SWR suggests that some power is being reflected back toward the transmitter, which can lead to signal loss and even potential damage to equipment. Using a directional wattmeter helps operators monitor SWR and ensure efficient operation.

What must be done to prevent standing waves on a feed line connected to an antenna? (G9A07)

A. The antenna feed point must be at DC ground potential
B. The feed line must be an odd number of electrical quarter wavelengths long
C. The feed line must be an even number of physical half-wavelengths long
D. The antenna feed point impedance must be matched to the characteristic impedance of the feed line

Standing waves on a feed line occur when the impedance of the antenna and the feed line mismatch. To prevent these standing waves and ensure maximum power is transferred from the transmitter to the antenna, the impedance at the feed point must be matched to the characteristic impedance of the feed line (typically 50 ohms in most amateur radio setups). Using an antenna tuner or adjusting the antenna design can help achieve this impedance match, reducing the formation of standing waves and improving overall system performance.

Preventing standing waves is critical to ensuring efficient signal transmission. The most effective solution is to match the antenna's feed point impedance with the feed line's impedance.

What causes reflected power at an antenna's feed point? (G9A04)

A. Operating an antenna at its resonant frequency
B. Using more transmitter power than the antenna can handle
C. A difference between feed line impedance and antenna feed point impedance
D. Feeding the antenna with unbalanced feed line

Reflected power occurs when the impedance at the antenna's feed point does not match the impedance of the feed line. This mismatch causes some of the power sent from the transmitter to be reflected back along the feed line instead of being radiated by the antenna. The greater the mismatch, the more energy is reflected, leading to less efficient signal transmission and potential damage to the transmitter over time.

Reflected power results from impedance mismatches between the antenna and feed line, reducing the efficiency of your transmission.

What is the effect of transmission line loss on SWR measured at the input to the line? (G9A11)

A. Higher loss reduces SWR measured at the input to the line
B. Higher loss increases SWR measured at the input to the line
C. Higher loss increases the accuracy of SWR measured at the input to the line
D. Transmission line loss does not affect the SWR measurement

---

Transmission line losses occur as the RF signal travels through the feed line, especially over longer distances or at higher frequencies. These higher losses can reduce the SWR value measured at the transmitter's end of the line, even if there is an impedance mismatch at the antenna. This occurs because the line losses attenuate both the forward and reflected signals, making the SWR appear lower than it really is at the feed point.

Transmission line loss can mask a high SWR at the antenna, making it necessary to measure SWR as close to the antenna as possible for accuracy.

SWR plays a crucial role in efficient signal transmission, and managing standing waves through impedance matching is essential to maintain signal strength. Impedance mismatches cause reflected power, reducing efficiency and risking equipment damage. Transmission line losses can complicate SWR measurements, which is why monitoring SWR at the feed point is vital.

SWR Measurement and Effects

This section delves into how SWR is measured, its effects on transmission line losses, and the relationships between feed line impedance and resistive loads.

What is the relationship between high standing wave ratio (SWR) and transmission line loss? (G9A02)

A. There is no relationship between transmission line loss and SWR
B. High SWR increases loss in a lossy transmission line
C. High SWR makes it difficult to measure transmission line loss
D. High SWR reduces the relative effect of transmission line loss

The high SWR indicates a mismatch between the transmission line and the antenna, causing power to reflect back toward the transmitter. This reflected power in a lossy transmission line leads to even more signal loss as it travels back and forth along the line. The greater the SWR, the more signal is lost, which can reduce the overall performance of your radio system. The higher the frequency, the more pronounced this effect becomes, which is why keeping SWR low is important for minimizing losses and improving transmissions.

A higher SWR reduces signal transmission and increases transmission line loss, especially in lossy coaxial cables.

If the SWR on an antenna feed line is 5:1, and a matching network at the transmitter end of the feed line is adjusted to present a 1:1 SWR to the transmitter, what is the resulting SWR on the feed line? (G9A08)

A. 1:1
B. 5:1
C. Between 1:1 and 5:1 depending on the characteristic impedance of the line
D. Between 1:1 and 5:1 depending on the reflected power at the transmitter

Even though a matching network can adjust the SWR to 1:1 at the transmitter end, the SWR on the feed line remains unchanged. The matching network helps the transmitter operate more efficiently by presenting it with the desired impedance. Still, the impedance mismatch between the antenna and feed line remains. As a result, the standing waves continue to exist on the feed line, and the SWR on the line remains at 5:1.

A matching network at the transmitter end does not change the SWR on the feed line itself; it only makes the transmitter "see" a 1:1 SWR.

What standing wave ratio results from connecting a 50-ohm feed line to a 200-ohm resistive load? (G9A09)

A. 4:1
B. 1:4
C. 2:1
D. 1:2

SWR is calculated as the ratio between the load and feed line impedance. In this case, the 200-ohm load is four times the 50-ohm feed line impedance, resulting in an SWR of 4:1. This means that a significant portion of the power is being reflected back toward the transmitter. The system is not operating at peak efficiency.

A mismatch between the feed line and the load results in a high SWR, such as 4:1 when connecting a 50-ohm feed line to a 200-ohm load.

What standing wave ratio results from connecting a 50-ohm feed line to a 10-ohm resistive load? (G9A10)

A. 2:1
B. 1:2
C. 1:5
D. 5:1

In this case, the load impedance is much lower than the feed line impedance, creating a greater mismatch. SWR is calculated as the higher impedance ratio to the lower impedance. Therefore, dividing 50 ohms by 10 ohms results in an SWR of 5:1. This significant mismatch means that most of the power is reflected back to the transmitter, making the system highly inefficient.

A large impedance mismatch, such as connecting a 50-ohm line to a 10-ohm load, results in a high SWR, in this case, 5:1. High SWR values lead to increased transmission line loss and reduced system performance while matching the feed line impedance to the antenna can minimize standing waves optimizing the system's overall performance.

Antenna Analyzer for SWR Measurements

When measuring SWR with an antenna analyzer, it is essential that both the antenna and feed line are connected to the analyzer. The analyzer sends signals through the feed line to the antenna and measures how much of that signal is reflected back. This helps the operator adjust the antenna or feed line for optimal performance. However, strong signals from nearby transmitters can interfere with SWR readings on an antenna analyzer. These signals, picked up by the analyzer, can distort the measurements, leading to inaccurate readings and making it harder to fine-tune the system.

Which of the following must be connected to an antenna analyzer when it is being used for SWR measurements? (G4B11)

A. Receiver
B. Transmitter
C. Antenna and feed line
D. All these choices are correct

What effect can strong signals from nearby transmitters have on an antenna analyzer? (G4B12)

A. Desensitization which can cause intermodulation products which interfere with impedance readings
B. Received power that interferes with SWR readings
C. Generation of harmonics which interfere with frequency readings
D. All these choices are correct

Tools like directional wattmeters and antenna analyzers help ham radio operators measure SWR, ensuring that power is efficiently transferred to the antenna.

Chapter Summary

Antennas are key to any ham radio station, converting electrical signals into radio waves. From simple wire antennas to dipoles and verticals, each type has unique characteristics that affect performance. Understanding feed point impedance, radiation patterns, and tuning helps operators optimize communication. Mastering these basics improves station efficiency and sets the stage for exploring more advanced antenna systems.

Chapter 15

Directional Antennas

A DIRECTIONAL ANTENNA IS a general term for any antenna that focuses its signal in a specific direction rather than radiating equally in all directions (like an omnidirectional antenna). This focus increases the signal strength in the desired direction, improving performance for transmitting and receiving signals over long distances. Common directional antennas include Yagi, log-periodic, and parabolic antennas.

Yagi Antennas

Yagi antennas are prized for their high gain and directionality, making them ideal for long-distance communication. With a driven element, reflector, and directors arranged on a boom, Yagis focus signals in one direction, boosting strength and improving reception. This efficiency makes them a favorite for amateur radio operators seeking reliable performance over long distances. However, their setup requires space, sturdy support, and often a rotator for precise aiming. Understanding key concepts like element length, front-to-back ratio, and radiation lobes is crucial for optimizing their performance and maximizing range.

What is the approximate length of the driven element of a Yagi antenna? (G9C02)

A. 1/4 wavelength
B. 1/2 wavelength
C. 3/4 wavelength
D. 1 wavelength

The driven element of a Yagi antenna is responsible for receiving and transmitting the signal, and it is typically around half the wavelength of the frequency it operates on. This length allows for maximum signal capture and efficient energy radiation, which is why the driven element is precisely tuned to this length for optimal performance. For example, if you're operating on a 20-meter band, the driven element will be about 10 meters long.

How do the lengths of a three-element Yagi reflector and director compare to that of the driven element? (G9C03)

A. The reflector is longer, and the director is shorter
B. The reflector is shorter, and the director is longer
C. They are all the same length
D. Relative length depends on the frequency of operation

In a Yagi antenna, the reflector is slightly longer and placed behind the driven element. It helps to reflect energy forward, increasing the forward gain. The director is shorter and positioned in front of the driven element to guide the signal in the desired direction. This combination creates a directional beam, which makes Yagi antennas highly effective in targeting specific areas while reducing interference from other directions.

What does "front-to-back ratio" mean in reference to a Yagi antenna? (G9C07)

A. The number of directors versus the number of reflectors
B. The relative position of the driven element with respect to the reflectors and directors
C. The power radiated in the major lobe compared to that in the opposite direction
D. The ratio of forward gain to dipole gain

The front-to-back ratio is a measurement that compares the power radiated in the Yagi antenna's main direction (front) to the power radiated in the opposite direction (back). A higher front-to-back ratio means that the antenna focuses more power in the desired direction and minimizes power radiated in reverse, improving signal quality by reducing unwanted noise and interference from behind the antenna.

What is meant by the "main lobe" of a directive antenna? (G9C08)

A. The magnitude of the maximum vertical angle of radiation
B. The point of maximum current in a radiating antenna element
C. The maximum voltage standing wave point on a radiating element
D. The direction of maximum radiated field strength from the antenna

---

The "main lobe" of a Yagi antenna refers to the direction in which the antenna radiates the most power. This is the area of maximum signal strength, where communication will be strongest and most reliable. For Yagi antennas, the main lobe is directed forward toward the director elements, allowing the antenna to focus its energy in that direction and improving its effectiveness for long-distance communication.

Yagi antennas are powerful tools for directional communication, allowing radio operators to focus their transmissions in a specific direction while reducing interference from other areas. Understanding the role of each element—the driven element, reflector, and director—along with concepts like front-to-back ratio and the main lobe is key to optimizing antenna performance. These principles help radio enthusiasts use Yagi antennas for more effective and targeted communication.

Enhancing Performance

Yagi antennas are well-known for their ability to focus energy in a specific direction, which makes them excellent for long-distance communication. However, depending on your needs, several ways to improve their performance exist. Factors such as element diameter, boom length, the number of directors, and the spacing between elements can significantly impact a Yagi's bandwidth, gain, and front-to-back ratio. These adjustments allow users to fine-tune their Yagi antennas for better communication performance in varying conditions.

Which of the following would increase the bandwidth of a Yagi antenna? (G9C01)

A. Larger-diameter elements
B. Closer element spacing
C. Loading coils in series with the element
D. Tapered-diameter elements

Increasing the diameter of the elements in a Yagi antenna broadens the range of frequencies over which the antenna can operate efficiently. This is because larger elements lower the antenna's "Q factor," making it less sensitive to small changes in frequency. A lower Q factor means a wider bandwidth, allowing the antenna to transmit and receive signals over a broader range of frequencies without a significant drop in performance. This can be particularly useful when dealing with varying frequencies or when working in environments with a lot of signal interference.

What is the primary effect of increasing boom length and adding directors to a Yagi antenna? (G9C05)

A. Gain increases
B. Beamwidth increases
C. Front-to-back ratio decreases
D. Resonant frequency is lower

By increasing the boom length and adding more directors to a Yagi antenna, you focus the antenna's energy more narrowly in a specific direction. This results in an increase in antenna gain, meaning more signal strength is directed forward. The trade-off is that a longer boom and additional directors can make the antenna more directional, narrowing the beamwidth. Still, this enhanced focus is beneficial for reaching distant stations with greater signal strength.

Which of the following can be adjusted to optimize forward gain, front-to-back ratio, or SWR bandwidth of a Yagi antenna? (G9C10)

A. The physical length of the boom
B. The number of elements on the boom
C. The spacing of each element along the boom
D. All these choices are correct

Several factors can be adjusted to optimize the performance of a Yagi antenna. Increasing the boom length and adding more elements can improve forward gain, while careful spacing of the elements can fine-tune the front-to-back ratio, which measures how well the antenna rejects signals from the opposite direction. By adjusting these physical parameters, operators can achieve better signal reception and transmission tailored to their specific needs, whether that is increasing signal strength, improving selectivity, or widening bandwidth.

Optimizing a Yagi antenna's performance involves manipulating various physical characteristics like element size, boom length, and the number and spacing of directors. These adjustments allow radio operators to enhance the antenna's bandwidth, increase its gain, and fine-tune its directional properties. Understanding these factors is key to maximizing a Yagi antenna's potential, ensuring strong, focused communication over long distances with minimal interference.

Matching Techniques and Advanced Configurations

Matching techniques are used to ensure maximum power transfer from your transmitter to your antenna by minimizing reflected power. Different methods, such as the beta or hairpin match and the gamma match, are commonly used with Yagi antennas and other advanced configurations. These techniques help fine-tune impedance mismatches that can otherwise decrease efficiency and performance. By using specific matching methods, operators can optimize their antenna system for better transmission and reception, even when dealing with complex antenna setups.

What is a beta or hairpin match? (G9C11)

A. A shorted transmission line stub placed at the feed point of a Yagi antenna to provide impedance matching
B. A 1/4 wavelength section of 75-ohm coax in series with the feed point of a Yagi to provide impedance matching
C. A series capacitor selected to cancel the inductive reactance of a folded dipole antenna
D. A section of a 300-ohm twin-lead transmission line used to match a folded dipole antenna

The beta or hairpin match is a clever solution for impedance matching in Yagi antennas. Essentially, it is a shorted transmission line stub placed at the antenna's feed point. This stub helps transform the feed point impedance to a more desirable value—typically matching it to the feed line's impedance. This ensures that more of the signal energy is transmitted effectively, reducing losses due to mismatch. The simplicity and effectiveness of the beta match make it a popular choice for fine-tuning Yagi antennas, especially in high-performance settings.

Which of the following is a characteristic of using a gamma match with a Yagi antenna? (G9C12)

A. It does not require the driven element to be insulated from the boom
B. It does not require any inductors or capacitors
C. It is useful for matching multiband antennas
D. All these choices are correct

Gamma matching is another impedance technique often used with Yagi antennas. One of its advantages is that it doesn't require the driven element to be insulated from the boom, which simplifies the antenna construction. The gamma match adjusts the impedance by using a gamma rod connected partway along the driven element. This method allows the operator to match the antenna's impedance to the feed line without significantly modifying the boom or driven element, making it both a practical and efficient option.

Matching techniques like the beta and gamma match are used to fine-tune antennas, especially Yagi antennas, to ensure they perform optimally. These methods allow for efficient impedance matching without significant modifications to the antenna structure, making them ideal for both casual operators and advanced users looking to maximize their communication system's effectiveness. Understanding these configurations helps operators achieve better signal transmission, improving overall radio performance.

Advanced Configurations and Gain Comparisons

When optimizing the performance of Yagi antennas, especially in advanced configurations, operators often aim for increased gain and better signal directivity. Techniques like stacking Yagi antennas and understanding gain measurements (dBd vs. dBi) play a key role in fine-tuning antenna systems. These configurations allow for stronger signals and improved range, which is critical for long-distance communication and competitive performance in amateur radio.

In free space, how does the gain of two three-element, horizontally polarized Yagi antennas spaced vertically 1/2 wavelength apart typically compare to the gain of a single three-element Yagi? (G9C09)

A. Approximately 1.5 dB higher
B. Approximately 3 dB higher
C. Approximately 6 dB higher
D. Approximately 9 dB higher

Stacking two Yagi antennas vertically with a separation of 1/2 wavelength is a popular method for increasing antenna gain. By aligning the antennas in this way, the overall system increases its directional power and efficiency, resulting in approximately a 3 dB increase in gain compared to a single Yagi antenna. This is significant because every 3 dB increase in gain effectively doubles the power radiated in the desired direction, allowing for improved signal strength and coverage. This technique is often used when higher performance is needed in competitive or long-range communications.

How does antenna gain in dBi compare to gain stated in dBd for the same antenna? (G9C04)

A. Gain in dBi is 2.15 dB lower
B. Gain in dBi is 2.15 dB higher
C. Gain in dBd is 1.25 dBd lower
D. Gain in dBd is 1.25 dBd higher

Antenna gain can be expressed in two ways: dBd (relative to a dipole) and dBi (relative to an isotropic radiator). A dipole is a real-world antenna, while an isotropic radiator is a theoretical antenna that radiates equally in all directions. Gain in dBi is always 2.15 dB higher than in dBd because a dipole antenna has a natural gain of 2.15 dB over an isotropic radiator. Understanding this distinction helps operators accurately compare antenna types and determine their performance in specific configurations.

Advanced configurations such as stacking Yagi antennas and knowing the difference between dBi and dBd gain measurements are crucial for maximizing antenna performance. These strategies allow operators to boost signal strength, improve range, and fine-tune their setups for specific communication needs. By leveraging these techniques, radio operators can significantly enhance their transmission power and ensure more reliable communication over greater distances.

Vertically Stacking

Understanding antenna configurations is crucial for optimizing signal strength and range in ham radio operations. Different configurations, such as vertically stacked Yagi antennas or log-periodic antennas, offer unique advantages that can be tailored to specific communication needs. This section explores how these configurations affect antennas' gain, directivity, and bandwidth, providing insights into the technical details that make them efficient tools for operators.

What is an advantage of vertically stacking horizontally polarized Yagi antennas? (G9D05)

A. It allows quick selection of vertical or horizontal polarization
B. It allows simultaneous vertical and horizontal polarization
C. It narrows the main lobe in azimuth
D. It narrows the main lobe in elevation

Vertically stacking Yagi antennas is a technique used to improve the performance of an antenna system. By stacking two or more horizontally polarized Yagis one above the other, the main lobe of the radiation pattern narrows in elevation. This enhances the antenna's gain in the desired direction, focusing the signal and reducing wasted energy in other directions. This is particularly useful in DX communication, where a focused signal can improve reception at the target destination, especially when communicating at low elevation angles.

Log Periodic Antenna

A log-periodic antenna is characterized by its unique design, where the lengths of its elements and their spacing along the boom vary according to a logarithmic scale.

Log-periodic antennas differ from Yagi antennas in that they are designed to provide consistent performance across a wide frequency range rather than being optimized for a single frequency. This makes log-periodic antennas ideal for multi-band operation. Their ability to resonate at multiple frequencies comes from their unique design, where element lengths and spacing vary in a logarithmic progression, enabling broad bandwidth and versatility for various communication needs.

Which of the following describes a log-periodic antenna? (G9D07)

A. Element length and spacing vary logarithmically along the boom
B. Impedance varies periodically as a function of frequency
C. Gain varies logarithmically as a function of frequency
D. SWR varies periodically as a function of boom length

Which of the following is an advantage of a log-periodic antenna? (G9D06)

A. Wide bandwidth
B. Higher gain per element than a Yagi antenna
C. Harmonic suppression
D. Polarization diversity

A log-periodic antenna is designed to offer a <u>wide operational bandwidth</u>, meaning it can function efficiently across a broad range of frequencies. This characteristic makes log-periodic antennas highly versatile, as they can be used without needing to retune the antenna for different frequencies. This makes them ideal for applications that require communication on multiple bands, such as in amateur radio or other professional communication systems where versatility is vital.

Long-Path Communication

In amateur radio, long-path communication refers to a signal traveling the long way around the Earth to reach its destination, typically covering a greater distance than the direct short-path route. Signals can travel in both directions around the globe, and sometimes, the long path offers better propagation conditions, especially when short-path signals are weak or obstructed. Understanding how to use short and long paths can improve a ham radio operator's ability to reach distant stations under varying conditions.

How is a directional antenna pointed when making a "long-path" contact with another station? (G2D06)

A. Toward the rising sun
B. Along the gray line
C. 180 degrees from the station's short-path heading
D. Toward the north

<u>When making a long-path contact, the operator points their directional antenna 180 degrees opposite from the short-path heading.</u> For example, if the short path to a station is to the east, the operator would point their antenna west to reach the same station via the long path. This technique takes advantage of global propagation conditions, such as ionospheric reflection,

that might favor long-path transmission at certain times of day or during specific solar conditions. Knowing when to use long-path communication can help an operator maintain reliable contacts even when short-path signals are poor.

Chapter Summary

Directional antennas, such as Yagi and log-periodic designs, are essential tools for amateur radio operators looking to focus their signals and maximize performance. By concentrating energy in specific directions, these antennas enhance gain, improve signal clarity, and reduce interference from unwanted sources. Yagi antennas are prized for their high gain and precision, making them ideal for long-distance (DX) communication, while log-periodic antennas provide broad bandwidth for seamless operation across multiple bands.

Key concepts like element arrangement, front-to-back ratio, and main lobes help operators optimize their Yagi antennas for efficiency and reliability. Advanced configurations, such as stacking Yagis or employing impedance-matching techniques like the gamma or beta match, allow for fine-tuning performance based on operating conditions. Meanwhile, the log-periodic antenna stands out for its versatility, handling multi-band operation without frequent adjustments.

By mastering the principles and configurations of directional antennas, amateur radio enthusiasts can improve signal strength, minimize interference, and achieve reliable communication over long distances. Whether chasing DX or ensuring steady connections across multiple bands, these antennas empower operators to get the most out of their stations.

Chapter 16

Specialized Antennas

SPECIALIZED ANTENNAS SERVE UNIQUE purposes that go beyond the standard designs like dipoles, Yagis, or verticals, catering to specific operational needs and environments. These antennas include loop antennas for compact setups, helical antennas for circular polarization, and beverage antennas for long-distance reception. Each type offers distinct advantages, such as improved performance in challenging conditions, portability, or specific polarization patterns, making them invaluable tools for operators with unique communication goals. Understanding these specialized antennas expands your versatility as an amateur radio operator, enabling you to tackle everything from low-noise reception to satellite communication with confidence.

Near Vertical Incidence Skywave Antenna

A Near Vertical Incidence Skywave (NVIS) antenna is a specialized antenna designed to achieve reliable short-range HF communication (typically within 300-500 miles) by utilizing skywave propagation at near-vertical angles. Unlike traditional long-distance antennas, which aim to send signals at low angles toward the horizon, an NVIS antenna directs most of its energy upward at steep angles. These signals are then reflected back to Earth by the ionosphere, covering a broad area around the transmitter.

Which of the following antenna types will be most effective as a near vertical incidence skywave (NVIS) antenna for short-skip communications on 40 meters during the day? (G9D01)

A. A horizontal dipole placed between 1/10 and 1/4 wavelength above the ground
B. A vertical antenna placed between 1/4 and 1/2 wavelength above the ground
C. A horizontal dipole placed at approximately 1/2 wavelength above the ground
D. A vertical dipole placed at approximately 1/2 wavelength above the ground

---

A horizontal dipole placed relatively close to the ground, between 1/10 and 1/4 wavelength, is ideal for Near Vertical Incidence Skywave (NVIS) propagation. NVIS is a propagation method used for short-skip communications, where radio waves are reflected almost vertically by the ionosphere, allowing them to travel short distances effectively. This type of setup is commonly used on the 40-meter band for reliable daytime communications over a range of a few hundred miles, making it valuable for local and regional communication in emergency networks or casual contact.

A NVIS antenna is ideal when you need reliable short-range HF communication in areas where traditional line-of-sight methods, like VHF/UHF, fail due to obstructions such as mountains, forests, or urban environments. This makes NVIS antennas particularly valuable in emergency situations, where regional communication is critical, such as coordinating disaster relief efforts or maintaining contact across rugged terrain. They are also useful in rural areas with sparse infrastructure, providing dependable communication over distances of 300-500 miles. Additionally, NVIS antennas are commonly used by the military and emergency services for tactical operations, as they excel in covering local areas without the need for repeaters or other infrastructure. Their low deployment height and ability to work effectively on HF bands make them a practical choice for portable and field operations in challenging environments.

Beverage Antenna

The Beverage antenna is a long-wire <u>directional receiving antenna primarily used for medium-frequency (MF) and low-high-frequency (HF) bands</u>. Its main advantage is its excellent directional characteristics, which make it ideal for long-distance reception in the direction the antenna is pointed. Because of its low noise and ability to capture weak signals, the Beverage antenna is popular among operators who focus on long-range DXing, especially on bands like 160 meters, where noise can be a significant problem.

What is the primary use of a Beverage antenna? (G9D09)

A. Directional receiving for MF and low HF bands
B. Directional transmitting for low HF bands
C. Portable direction finding at higher HF frequencies
D. Portable direction finding at lower HF frequencies

The Beverage antenna is a specialized tool for amateur radio operators and shortwave listeners who prioritize weak signal reception, especially on the low bands. Its simplicity, effectiveness, and high SNR make it a favorite for low-frequency DXing and challenging listening environments.

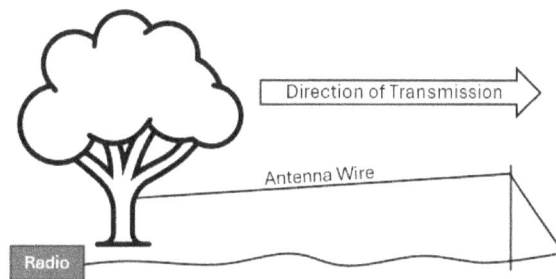

Halo Antenna

Understanding antenna directionality and radiation patterns is important for maximizing a radio station's performance. Some antennas, like the VHF/UHF "halo" antenna, are designed to radiate equally in all directions, while others, such as electrically small loop antennas, have specific directions of nulls or weaker signals. Knowing how these radiation patterns affect signal transmission and reception helps operators choose a suitable antenna for their needs, whether for local omnidirectional communication or long-distance directional contact.

In which direction is the maximum radiation from a VHF/UHF "halo" antenna? (G9D03)

A. Broadside to the plane of the halo
B. Opposite the feed point
C. Omnidirectional in the plane of the halo
D. On the same side as the feed point

A "halo" antenna is a horizontally polarized antenna, often used for VHF/UHF communications, with an <u>omnidirectional radiation pattern in the halo plane.</u> This means it radiates signals equally in all directions around the antenna, making it ideal for mobile or portable applications where communication is needed in all directions without requiring constant antenna repositioning. However, the signal strength is strongest in the horizontal plane, while the vertical radiation (above and below the antenna) is weaker, which aligns well with typical communication scenarios on these bands.

Loop Antenna

A loop antenna is a type of antenna formed by a loop or coil of conductive wire that is often used for both transmitting and receiving signals. Loop antennas are versatile and come in various sizes and designs, such as small magnetic loops for receiving or large full-wavelength loops for transmitting. Their compact size, directional properties, and efficiency make them popular among amateur radio operators, shortwave listeners, and professionals in RF communication.

In which direction or directions does an electrically small loop (less than 1/10 wavelength in circumference) have nulls in its radiation pattern? (G9D10)

A. In the plane of the loop
B. Broadside to the loop
C. Broadside and in the plane of the loop
D. Electrically small loops are omnidirectional

An electrically small loop antenna, one whose circumference is less than 1/10 of a wavelength, has a characteristic radiation pattern with nulls, or points of minimal radiation, <u>broadside to the loop</u>. In simpler terms, the weakest signals are emitted directly perpendicular to the plane of the loop. This makes small loop antennas useful for direction-finding applications because the nulls can be used to pinpoint the direction of incoming signals by rotating the antenna until the signal fades. It also means that maximum radiation occurs along the axis of the loop, making it an effective antenna for specific directional applications.

Loop antennas offer a compact and efficient solution for operators seeking low-noise reception and stealth installations. These antennas can be magnetic loops or large wire loops, each with distinct characteristics. Magnetic loop antennas are particularly noted for their compact size and low noise reception, making them suitable for use in noisy environments. They are also highly portable, ideal for field operations or situations where a full-sized antenna is impractical. However, loop antennas generally have lower radiation efficiency and require precise tuning to operate effectively. This complexity in tuning can be a drawback for some operators.

Mobile Antenna

Mobile antennas are specifically designed for use on vehicles, enabling amateur radio operators to stay connected while on the move. These antennas are compact and rugged, built to withstand the challenges of mobile environments, such as vibration, weather exposure, and limited mounting space. Despite their smaller size compared to base station antennas, mobile antennas are engineered for efficient performance on popular bands, often using techniques like loading coils to achieve resonance. Whether you're engaging in local communication or participating in HF activities while traveling, mobile antennas are a critical part of ensuring reliable and flexible radio operation on the road.

How does a "screwdriver" mobile antenna adjust its feed point impedance? (G9D08)

A. By varying its body capacitance
B. By varying the base loading inductance
C. By extending and retracting the whip
D. By deploying a capacitance hat

A "screwdriver" mobile antenna is a type of adjustable antenna designed for use in vehicles, offering flexibility across different frequency bands. It adjusts its feed point impedance by <u>varying the base loading inductance.</u> This is accomplished by moving a coil or inductor up and down, effectively changing the antenna's electrical length. This feature allows the operator to tune the antenna across multiple frequency bands without needing to stop and manually adjust it, making it highly convenient for mobile ham radio operators who need quick and efficient impedance adjustments.

Trapped Antenna

Trapped antennas are versatile and efficient solutions for multi-band operation, designed to work seamlessly across multiple frequencies without requiring manual adjustments. These antennas use traps—coils and capacitors arranged in parallel—that act as resonant circuits to isolate specific sections of the antenna at different frequencies. By effectively "shortening" the antenna at higher frequencies while using its full length at lower ones, trapped antennas enable operators to cover multiple bands with a single design. Their convenience and simplicity make them a popular choice for those seeking efficient multi-band performance in limited space.

What is the primary function of antenna traps? (G9D04)

A. To enable multiband operation
B. To notch spurious frequencies
C. To provide balanced feed point impedance
D. To prevent out-of-band operation

Antenna traps are a key component used to enable multiband operation in antennas. These are resonant circuits that are placed along the length of an antenna, allowing it <u>to operate efficiently on multiple bands</u> without needing separate antennas for each band. When transmitting on a specific band, the trap effectively isolates parts of the antenna, allowing only the relevant portion to radiate the signal. This allows the same antenna to work across different frequency ranges, offering versatility and reducing the need for multiple antennas in a compact setup.

Antenna Wire

Traps   Traps

Coaxial Cable
to Radio

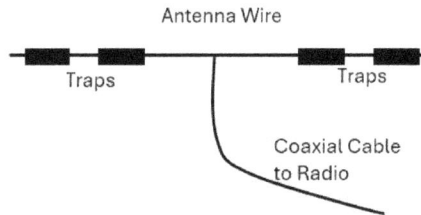

Multi-Band Antenna

A multi-band antenna is an antenna designed to operate efficiently on multiple frequency bands without requiring significant adjustments or additional components. These antennas allow amateur radio operators to communicate across various bands, making them versatile and convenient for both HF and VHF/UHF operation.

Which of the following is a disadvantage of multiband antennas? (G9D11)

A. They present low impedance on all design frequencies
B. They must be used with an antenna tuner
C. They must be fed with open wire line
D. They have poor harmonic rejection

While multiband antennas offer the convenience of operating on multiple frequency bands with a single antenna, one disadvantage is that they tend to have poor harmonic rejection. Harmonics are unwanted signals that are multiples of the fundamental frequency, and they can cause interference or reduce the clarity of communication. Due to their design, multiband antennas may allow these harmonics to radiate more easily. This can be problematic in environments where clear and precise signal transmission is crucial.

Multi-band antennas are an excellent choice for those who need flexibility and versatility in their setups. They offer reliable performance across multiple frequency ranges with minimal complexity.

Chapter Summary

Specialized antennas expand the capabilities of amateur radio operators, addressing unique challenges and requirements that standard antennas cannot. From the versatile NVIS antenna for short-range HF communication to the noise-reducing Beverage antenna for long-distance reception, each type serves a specific purpose. Loop antennas offer compact and efficient solutions for low-noise environments, while mobile and trapped antennas bring convenience to operators on the move or with limited space. Multi-band antennas add flexibility, enabling operation across multiple frequencies with a single setup.

By understanding the design, operation, and applications of these specialized antennas, operators can tailor their stations to achieve optimal performance in a variety of scenarios. Whether tackling challenging environments, enhancing portability, or maximizing versatility, these antennas empower amateur radio enthusiasts to broaden their horizons and elevate their communication capabilities.

Chapter 17

Power

Solar Energy

Solar power is an increasingly popular energy solution for ham operators.

In a typical ham radio solar setup, solar panels are connected to batteries through a charge controller to manage power storage. This enables the station to continue operating even when the sun is not shining. Key components such as photovoltaic cells, charge controllers, and series diodes work together to ensure that energy is harvested and stored safely and efficiently. By understanding the technical aspects of solar panel configurations, voltages, and protection mechanisms, radio operators can maintain reliable power systems, even in challenging environments.

This section will delve into the essentials of solar power, including proper setup, precautions, and the benefits of using solar energy for ham radio operations.

Solar Panel Configurations and Precautions

Individual photovoltaic (PV) cells are typically connected in a series-parallel configuration in solar panels. This setup combines the benefits of both series and parallel wiring. When cells are connected in series, their voltages add up, which helps produce a higher overall voltage. When connected in parallel, their currents add up, providing a higher current output. This configuration ensures that a solar panel can generate enough voltage and current to be useful for charging batteries or powering equipment, making it ideal for various applications, including setups in off-grid environments.

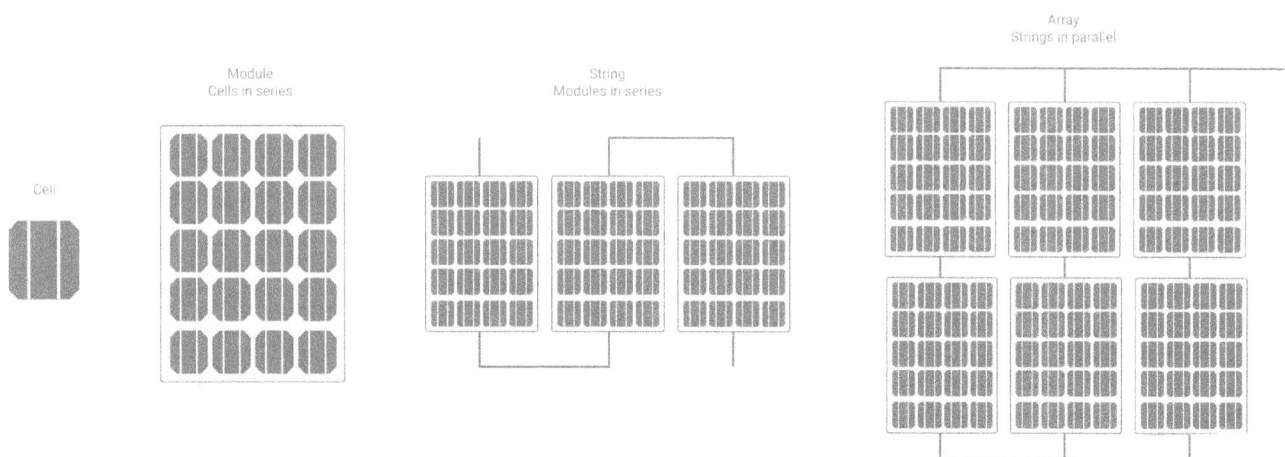

Cell

Module
Cells in series

String
Modules in series

Array
Strings in parallel

In what configuration are the individual cells in a solar panel connected together? (G4E08)

A. Series-parallel
B. Shunt
C. Bypass
D. Full-wave bridge

It is vital to use a charge controller when using a solar panel to charge a lithium-iron-phosphate battery. The charge controller regulates the voltage and current coming from the solar panel to prevent overcharging the battery, which could damage it or reduce its lifespan. Lithium iron phosphate batteries have specific charging requirements, and a charge controller helps manage these conditions, ensuring the battery is charged efficiently and safely without being overcharged.

What precaution should be taken when connecting a solar panel to a lithium iron phosphate battery? (G4E11)

A. Ground the solar panel outer metal framework
B. Ensure the battery is placed terminals-up
C. A series resistor must be in place
D. The solar panel must have a charge controller

Solar Panel Voltages and Protection

When fully illuminated, a single silicon photovoltaic (PV) cell generates an open-circuit voltage of approximately 0.5 volts DC (VDC). This measurement represents the maximum voltage the cell can produce when no current is flowing—essentially, when the circuit is open. The open-circuit voltage is a critical parameter for understanding the performance of a PV cell and designing solar power systems.

What is the approximate open-circuit voltage from a fully illuminated silicon photovoltaic cell? (G4E09)

A. 0.02 VDC
B. 0.5 VDC
C. 0.2 VDC
D. 1.38 VDC

Multiple PV cells are connected in series to achieve higher voltages suitable for practical applications. When connected this way, the voltages of individual cells are added together. However, it's important to note that the actual output voltage under load conditions (when current flows) will be slightly lower due to factors like internal resistance and power draw.

Diodes for Protection

A series diode is necessary when connecting a solar panel to a storage battery. The diode acts as a one-way valve, allowing current to flow from the solar panel to the battery during daylight while preventing the reverse flow of current at night or during low illumination. Without a diode, the battery could discharge through the panel, reducing its efficiency and potentially damaging the battery and solar panels over time.

Why should a series diode be connected between a solar panel and a storage battery that is being charged by the panel? (G4E10)

A. To prevent overload by regulating the charging voltage
B. To prevent discharge of the battery through the panel during times of low or no illumination
C. To limit the current flowing from the panel to a safe value
D. To prevent damage to the battery due to excessive voltage at high illumination levels

Solar panels are configured and protected to ensure they work efficiently and safely in various applications. Proper configurations and protective devices like diodes and charge controllers are crucial for maintaining performance and longevity, especially when using lithium iron phosphate batteries or other sensitive equipment.

Rectifiers

Rectifiers are essential components in power supply circuits. Their primary function is to convert alternating current into direct current, a necessary process for powering most electronic devices. Understanding how different rectifier types function allows radio operators to design power supplies that provide clean, consistent power. In this section, we'll explore full-wave and half-wave rectifiers, their characteristics, and how they impact the conversion of AC to DC in radio equipment.

What portion of the AC cycle is converted to DC by a half-wave rectifier? (G7A05)

A. 90 degrees
B. 180 degrees
C. 270 degrees
D. 360 degrees

In a half-wave rectifier, only one-half of the AC cycle (either the positive or the negative half) is used to produce DC. This means only 180 degrees of the entire 360-degree AC waveform are converted into DC. During the other half of the cycle, no current flows. As a result, the output is less efficient and more "pulsed," creating a rougher DC signal that often requires additional filtering to be usable in sensitive circuits.

What is characteristic of a half-wave rectifier in a power supply? (G7A04)

A. Only one diode is required
B. The ripple frequency is twice that of a full-wave rectifier
C. More current can be drawn from the half-wave rectifier
D. The output voltage is two times the peak input voltage

A half-wave rectifier is simpler in design than a full-wave rectifier because it requires only one diode to block the negative portion of the AC signal, allowing only the positive half to pass through as DC. While this simplicity is an advantage in low-cost or low-power applications, the output is less efficient, with more ripple and less smooth DC than a full-wave rectifier.

Full-Wave Rectifiers

A full-wave rectifier uses two diodes along with a center-tapped transformer to convert the entire AC waveform (both positive and negative halves) into a usable DC output. The center tap allows the circuit to direct both halves of the AC cycle through the diodes, effectively doubling the frequency of the resulting DC pulses. This setup provides smoother power conversion than a half-wave rectifier, as it utilizes the full 360 degrees of the AC cycle. Full-wave rectifiers are more efficient and often used in higher-power applications requiring smoother DC.

Which type of rectifier circuit uses two diodes and a center-tapped transformer? (G7A03)

A. Full-wave
B. Full-wave bridge
C. Half-wave
D. Synchronous

A full-wave rectifier converts both the positive and negative halves of the AC cycle into DC, meaning it utilizes the entire 360-degree cycle. Compared to a half-wave rectifier, this results in a more continuous flow of current and less ripple in the output. The full-wave design ensures that the power supply delivers a more stable and efficient DC signal, which is particularly important for sensitive electronic equipment like radios.

What portion of the AC cycle is converted to DC by a full-wave rectifier? (G7A06)

A. 90 degrees
B. 180 degrees
C. 270 degrees
D. 360 degrees

When a full-wave rectifier is used without any filtering, the DC output is a series of pulses. These pulses occur at twice the frequency of the input AC signal because the full-wave rectifier inverts the negative half of the AC cycle, creating a positive pulse. This creates a pulsed DC signal, which can be smoothed using capacitors or inductors to create a more stable and continuous DC supply.

What is the output waveform of an unfiltered full-wave rectifier connected to a resistive load? (G7A07)

A. A series of DC pulses at twice the frequency of the AC input
B. A series of DC pulses at the same frequency as the AC input
C. A sine wave at half the frequency of the AC input
D. A steady DC voltage

Rectifiers are the backbone of power conversion in electronic systems, transforming AC into DC power. Understanding the difference between half-wave and full-wave rectification and the role of diodes and transformers helps ensure that the correct rectifier is chosen for specific applications. Full-wave rectifiers offer greater efficiency, while half-wave rectifiers are simpler but less effective. By mastering these concepts, operators can design more reliable power supplies that meet the needs of their radio equipment.

Power Supply

Switchmode

Power supplies are components that convert electrical energy into a form suitable for operating electronic devices. Two popular types of power supplies are switchmode and linear. Each has unique characteristics that make it ideal for different applications. Switchmode power supplies, known for their efficiency, utilize high-frequency operation, while linear power supplies offer a simpler design and lower noise but are generally less efficient. This section focuses on understanding the key differences between these two power supplies and how they impact performance and component selection in radio equipment.

Which of the following is characteristic of a switchmode power supply as compared to a linear power supply? (G7A08)

A. Faster switching time makes higher output voltage possible
B. Fewer circuit components are required
C. High-frequency operation allows the use of smaller components
D. Inherently more stable

Switchmode power supplies are highly efficient because they convert power by rapidly switching transistors on and off at high frequencies. This high-frequency operation allows the use of smaller components, such as transformers and capacitors, which reduces the overall size and weight of the power supply. Switchmode supplies also tend to generate less heat than linear supplies, making them more energy-efficient. However, because they operate at high frequencies, they can introduce more electrical noise, which may interfere with sensitive equipment unless adequately filtered.

Choosing between a switchmode and linear power supply depends on your equipment's needs. Switchmode power supplies offer significant efficiency and size advantages, particularly when high-frequency operation is necessary. However, they require proper noise filtering to avoid interference. Understanding these differences helps operators select the proper power supply for their radios, ensuring reliable and efficient performance in their setups.

Filtering

In a power supply, capacitors and inductors are components for smoothing out the ripple in the DC output after rectification. Capacitors store electrical energy and help stabilize voltage by filling in gaps in the DC signal while inductors resist changes in current. Together, they form a filter network that reduces unwanted AC ripple, ensuring the output is as close to pure DC as possible. This filtered DC is crucial for powering sensitive radio circuits without introducing noise or instability.

Which of the following components are used in a power supply filter network? (G7A02)

A. Diodes
B. Transformers and transducers
C. Capacitors and inductors
D. All these choices are correct

Protection

A power supply bleeder resistor is a safety component that ensures the capacitors in a power supply discharge their stored energy when the power is turned off. Capacitors can retain a charge for some time after the power is disconnected, which could be dangerous during maintenance or repair. The bleeder resistor slowly drains this stored energy, ensuring the system is safe to handle. Without a bleeder resistor, capacitors could deliver a shock if touched even after the equipment is turned off.

What is the function of a power supply bleeder resistor? (G7A01)

A. It acts as a fuse for excess voltage
B. It discharges the filter capacitors when power is removed
C. It removes shock hazards from the induction coils
D. It eliminates ground loop current

Chapter Summary

Powering your ham radio station effectively and safely is crucial for optimal performance, and understanding the diverse range of power sources and their components is key. From choosing between switchmode and linear power supplies to harnessing renewable energy like solar power, this chapter has equipped you with the knowledge to make informed decisions. We've explored the intricacies of rectifiers, filters, and power protection mechanisms, ensuring clean and stable energy for your setup. By mastering these concepts, you'll be well-prepared to maintain a reliable and efficient station, empowering your communication endeavors across all conditions.

Chapter 18

Electrical Safety

SAFETY HAS ALWAYS BEEN a cornerstone of amateur radio, emphasized heavily in the Technician's exam. Safety becomes even more critical as you progress to General Class, especially as you work with higher voltages, currents, and advanced setups. Electrical safety isn't just a precaution—protecting yourself, your equipment, and your surroundings is essential.

In ham radio, equipment often operates on mains electricity, presenting serious risks such as electric shock and potential fires if mishandled. Proper grounding is a cornerstone of safety, not only protecting your gear from electrical surges and static buildup but also reducing interference and ensuring a common electrical potential across all equipment. Using ground rods, copper wire, and lightning arrestors safeguards your station against these dangers.

Safe operation extends beyond grounding. Proper ventilation prevents overheating, fuses and circuit breakers protect against overloads, and a clean, organized workspace minimizes hazards. By following these best practices, you can enjoy amateur radio confidently and safely, knowing your setup is secure.

Power Safety

Emergency Generator Safety

Emergency generators provide backup power during outages; but they must be installed and operated safely. The primary concern is ensuring the generator is used in a well-ventilated area to prevent the buildup of dangerous exhaust gases like carbon monoxide. Carbon monoxide poisoning can occur quickly and without warning, so following safety guidelines is critical.

Which of the following is true of an emergency generator installation? (G0B09)

A. The generator should be operated in a well-ventilated area
B. The generator must be insulated from ground
C. Fuel should be stored near the generator for rapid refueling in case of an emergency
D. All these choices are correct

A generator should always be operated outdoors or in a well-ventilated area, never in enclosed spaces such as garages or basements. This prevents the accumulation of carbon monoxide, an odorless and lethal gas. By following this simple precaution, you can ensure that your generator provides power safely during an emergency without endangering the lives of those nearby.

Power Supply Interlock

Power supply interlocks are safety features designed to prevent accidental contact with dangerously high voltages. They ensure that the power is automatically disconnected when a device's cabinet is opened to avoid electrical shocks.

What is the purpose of a power supply interlock? (G0B12)

A. To prevent unauthorized changes to the circuit that would void the manufacturer's warranty
B. To shut down the unit if it becomes too hot
C. To ensure that dangerous voltages are removed if the cabinet is opened
D. To shut off the power supply if too much voltage is produced

A power supply interlock removes dangerous voltages when a power supply cabinet or similar device is opened. This safety mechanism prevents accidental exposure to high-voltage components, protecting operators from electric shock or injury. Always ensure that devices with interlocks are adequately maintained and functional to ensure safety during servicing.

National Electrical Code

Understanding the National Electrical Code (NEC) and Station Safety

The National Electrical Code (NEC) is a set of regulations and guidelines designed to ensure electrical safety in various installations, including amateur radio stations. Compliance with the NEC helps reduce the risk of electrical hazards, such as short circuits, overloads, and electrical fires. For ham radio operators, following these standards ensures that all electrical wiring, circuits, and connections in their station are safe and meet the required electrical codes. This is essential for personal safety and avoiding violations that could cause legal issues or lead to accidents.

Which of the following is covered by the National Electrical Code? (G0B06)

A. Acceptable bandwidth limits
B. Acceptable modulation limits
C. Electrical safety of the station
D. RF exposure limits of the human body

Minimum Wire Size for a 20-Amp Circuit (AWG 12)

The National Electrical Code specifies the minimum wire size for different circuit capacities to prevent overheating and fire hazards. For a 20-amp circuit, the minimum wire size allowed is AWG number 12. This gauge of wire is thick enough to handle the current safely without excessive heat buildup, ensuring that the wiring remains safe for prolonged use under load. Understanding wire sizes is crucial in amateur radio setups, especially when high-power equipment is used, as incorrect wiring can lead to failures or dangerous conditions.

According to the National Electrical Code, what is the minimum wire size that may be used safely for wiring with a 20-ampere circuit breaker? (G0B02)

A. AWG number 20
B. AWG number 16
C. AWG number 12
D. AWG number 8

AWG stands for American Wire Gauge, a standardized system used in North America to measure the thickness or diameter of electrically conductive wire. The AWG system assigns a number to the wire gauge, where higher numbers represent thinner wire and lower numbers represent thicker wire. For example, AWG 12 wire is thicker than AWG 18 wire.

The thickness of the wire affects its ability to carry current safely. Thicker wires (with lower AWG numbers) can carry more current without overheating, while thinner wires (with higher AWG numbers) are suitable for lower currents. This is important in electrical installations, including amateur radio setups, where different types of equipment may require wires of varying thicknesses to ensure safe operation.

Appropriate Fuse or Circuit Breaker Size for AWG 14 Wiring

In a circuit using AWG number 14 wiring, the maximum appropriate fuse or circuit breaker size is 15 amperes. This rating is selected to ensure that the wiring is not overloaded beyond its capacity, protecting the circuit from excessive current that could cause overheating. Using a fuse or breaker with a higher rating could allow dangerous current levels to flow through the wiring, leading to overheating or even fire. For amateur radio operators, properly sizing circuit protection devices is essential to maintain a safe operating environment.

Which size of fuse or circuit breaker would be appropriate to use with a circuit that uses AWG number 14 wiring? (G0B03)

A. 30 amperes
B. 25 amperes
C. 20 amperes
D. 15 amperes

Hot Wires and Circuit Protection in a 240 VAC Circuit

In a 240 VAC circuit with four conductors, only the hot wires should be connected to fuses or circuit breakers. This ensures that if there is an overload or short circuit, the current flowing through the hot wires is interrupted, preventing damage to equipment or the risk of electrical shock. The neutral and ground wires should not be connected to fuses or breakers, as they are not intended to carry live current under normal operation. This understanding is important in maintaining safety when dealing with high-voltage equipment.

Which wire or wires in a four-conductor 240 VAC circuit should be attached to fuses or circuit breakers? (G0B01)

A. Only the hot wires
B. Only the neutral wire
C. Only the ground wire
D. All wires

RF Safety

Amateur radio operators must ensure compliance with RF safety regulations to protect themselves and others from potential harm caused by radio frequency (RF) exposure. One key step is performing a routine RF exposure evaluation. This process assesses the amount of RF energy emitted by your station to determine whether it exceeds the Maximum Permissible Exposure (MPE) limits. If the evaluation reveals high-exposure areas, steps must be taken to prevent access to these areas, such as repositioning antennas, lowering power, or creating physical barriers.

Which of the following steps must an amateur operator take to ensure compliance with RF safety regulations? (G0A08)

A. Post a copy of FCC Part 97.13 in the station
B. Notify neighbors within a 100-foot radius of the antenna of the existence of the station and power levels
C. Perform a routine RF exposure evaluation and prevent access to any identified high exposure areas
D. All these choices are correct

RF Field Measurement

Operators use a calibrated field strength meter with a calibrated antenna to accurately measure RF field strength. This instrument helps measure the intensity of RF energy around the station, ensuring it remains within safe exposure limits. It is a crucial tool for verifying compliance with RF safety standards and identifying any concerns around the antenna.

What type of instrument can be used to accurately measure an RF field strength? (G0A09)

A. A receiver with digital signal processing (DSP) noise reduction
B. A calibrated field strength meter with a calibrated antenna
C. An SWR meter with a peak-reading function
D. An oscilloscope with a high-stability crystal marker generator

Actions for Excessive RF Exposure

Suppose an evaluation shows that RF energy from your station exceeds permissible limits for human absorption. In that case, you are required to take corrective action to prevent exposure. This might include limiting power output, increasing the distance between antennas and people, or adding warning signs. Failure to take such steps could result in harmful exposure to RF energy, which is regulated to prevent potential health risks.

What must you do if an evaluation of your station shows that the RF energy radiated by your station exceeds permissible limits for possible human absorption? (G0A05)

A. Take action to prevent human exposure to the excessive RF fields
B. File an Environmental Impact Statement (EIS-97) with the FCC
C. Secure written permission from your neighbors to operate above the controlled MPE limits
D. All these choices are correct

RF Exposure

Understanding RF Exposure and FCC Compliance

What stations are subject to the FCC rules on RF exposure? (G0A12)

A. All commercial stations; amateur radio stations are exempt
B. Only stations with antennas lower than one wavelength above the ground
C. Only stations transmitting more than 500 watts PEP
D. All stations with a time-averaged transmission of more than one milliwatt

RF exposure regulations ensure that amateur radio stations operate safely by preventing excessive exposure to radio frequency energy. According to the FCC rules, all stations transmitting more than one milliwatt of time-averaged power are subject to these RF exposure rules. This threshold applies broadly, covering most amateur radio stations, even those using relatively low power. Operators need to be aware of their station's power levels to ensure they comply with these regulations and minimize any potential risks associated with RF exposure.

If your station fails to meet the FCC's RF exposure exemption criteria, you are required to perform an RF Exposure Evaluation. This can be done following the FCC's OET Bulletin 65 guidelines, which provide detailed procedures for assessing RF safety. This evaluation ensures that your transmissions do not exceed the permissible limits for human exposure, protecting both yourself and others in the vicinity.

What must you do if your station fails to meet the FCC RF exposure exemption criteria? (G0A06)

A. Perform an RF Exposure Evaluation in accordance with FCC OET Bulletin 65
B. Contact the FCC for permission to transmit
C. Perform an RF exposure evaluation in accordance with World Meteorological Organization guidelines
D. Use an FCC-approved band-pass filter

How to Ensure Compliance with RF Exposure Regulations

Multiple ways exist to ensure that your station is FCC-compliant regarding RF exposure. These include calculations based on FCC OET Bulletin 65, computer modeling, or direct field strength measurements with calibrated equipment. By following these methods, you can verify that your station's RF emissions are within safe limits and avoid potential violations of FCC rules. Regular evaluations are necessary, especially when changing equipment or antenna setups.

How can you determine that your station complies with FCC RF exposure regulations? (G0A03)

A. By calculation based on FCC OET Bulletin 65
B. By calculation based on computer modeling
C. By measurement of field strength using calibrated equipment
D. All these choices are correct

RF Safety and Directional Antennas

When using a directional antenna, considering the potential for RF exposure to yourself and others, such as neighbors, is important. If an evaluation shows that your neighbor might experience more than the allowable RF exposure limit from the main lobe of your directional antenna, action must be taken. The recommended precaution is to ensure that the antenna cannot be pointed in their direction when they are present. This can involve physical restrictions on the rotation of the antenna or simply timing your transmissions when you know the area is clear.

What should be done if evaluation shows that a neighbor might experience more than the allowable limit of RF exposure from the main lobe of a directional antenna? (G0A10)

A. Change to a non-polarized antenna with higher gain
B. Use an antenna with a higher front-to-back ratio
C. Take precautions to ensure that the antenna cannot be pointed in their direction when they are present
D. All these choices are correct

Managing RF Exposure Risks

The FCC regulates RF exposure to protect individuals from excessive RF radiation, which can cause harmful effects like tissue heating. Directional antennas can concentrate RF energy in specific directions, potentially exposing people in those areas to higher-than-permissible levels. Therefore, amateur radio operators need to carefully evaluate and manage their antenna orientation to prevent unintentional exposure to others.

One of the most significant effects of RF energy on the human body is its ability to heat body tissue. Just like a microwave heats food, RF energy can cause tissue to heat up, potentially leading to burns or other damage if exposure is high or prolonged. This is why ensuring that exposure stays within regulated limits is critical for safety.

What is one way that RF energy can affect human body tissue? (G0A01)

A. It heats body tissue
B. It causes radiation poisoning
C. It causes the blood count to reach a dangerously low level
D. It cools body tissue

Understanding RF Exposure from a Transmitted Signal

Several factors must be considered to determine RF exposure from a transmitted signal. These include duty cycle, frequency, and power density. The duty cycle refers to the proportion of time the transmitter is actively sending a signal. For example, continuous transmissions have a higher duty cycle compared to intermittent transmissions. Frequency plays a role because RF exposure limits vary with different frequencies, as higher frequencies are typically more likely to penetrate human tissues. Power density, or the amount of RF energy in a given area, is another key factor, as higher power densities can increase exposure risk. Together, these factors help ensure that RF exposure stays within safe limits.

Which of the following is used to determine RF exposure from a transmitted signal? (G0A02)

A. Its duty cycle
B. Its frequency
C. Its power density
D. All these choices are correct

Modulation Duty Cycle and RF Exposure

The modulation duty cycle of a transmission also impacts RF exposure. <u>A lower duty cycle allows greater power levels to be transmitted while maintaining safe exposure limits.</u> Lower duty cycles mean the transmitter is off for a portion of the time, reducing overall exposure. Understanding how the duty cycle interacts with power levels helps operators maximize transmission efficiency while maintaining compliance with safety regulations.

What is the effect of modulation duty cycle on RF exposure? (G0A07)

A. A lower duty cycle permits greater power levels to be transmitted
B. A higher duty cycle permits greater power levels to be transmitted
C. Low duty cycle transmitters are exempt from RF exposure evaluation requirements
D. High duty cycle transmitters are exempt from RF exposure requirements

---

Time Averaging in RF Radiation Exposure

<u>Time averaging refers to evaluating RF radiation exposure over a defined period</u>, rather than at a single instant. This approach smooths out fluctuations in exposure levels, considering both high and low power transmission times. For example, even if a high-power transmission occurs for a short time, if it is followed by periods of low-power or no transmission, the time-averaged exposure may remain within safe limits. This concept is essential for accurately assessing compliance with FCC guidelines, particularly for variable transmission power operations.

What does "time averaging" mean when evaluating RF radiation exposure? (G0A04)

A. The average amount of power developed by the transmitter over a specific 24-hour period
B. The average time it takes RF radiation to have any long-term effect on the body
C. The total time of the exposure
D. The total RF exposure averaged over a certain period

Grounding & Bonding

Proper grounding and bonding are vital for both safety and performance. Grounding dissipates electrical surges, prevents static buildup, and shields equipment from lightning-induced damage. Ground rods should be driven deep into moist soil, and copper wire should connect all equipment to a common ground point. Bonding ensures all components share the same electrical potential, reducing the risk of shock and interference. Installing lightning arrestors adds an extra layer of protection, safeguarding your setup from harmful surges.

Grounding

Grounding refers to connecting the equipment to the earth, typically through a ground rod. Grounding provides a safe path for electrical energy to dissipate into the ground, like lightning strikes or electrical surges. The primary goal of grounding is safety, ensuring that any excess electrical charge doesn't remain on equipment and potentially harm operators.

Grounding all metal enclosures of station equipment is critical for safety. <u>The reason behind this is simple: it ensures that any hazardous voltages, such as those from power surges or faulty wiring, cannot build up on your equipment's chassis or metal surfaces.</u> If equipment isn't grounded correctly, there's a risk that dangerous voltages might appear on the metal case, which could lead to electrical shocks if touched. Proper grounding provides a path for those stray voltages to safely dissipate into the earth, reducing the risk of electrical hazards and improving the station's overall performance by reducing interference and hum.

Why must all metal enclosures of station equipment be grounded? (G4C12)

A. It prevents a blown fuse in the event of an internal short circuit
B. It prevents signal overload
C. It ensures that the neutral wire is grounded
D. It ensures that hazardous voltages cannot appear on the chassis

Consequences of Incorrect Grounding

What is a possible cause of high voltages that produce RF burns? (G4C05)

A. Flat braid rather than round wire has been used for the ground wire
B. Insulated wire has been used for the ground wire
C. The ground rod is resonant
D. The ground wire has high impedance on that frequency

A ground wire with high impedance can cause high voltages that produce RF burns at a specific frequency. When the ground wire has a high impedance, it does not effectively carry RF energy to the ground, allowing high voltage to accumulate on the surfaces of equipment enclosures. This can result in RF burns if you touch the equipment, especially at points where high RF voltage is present. The solution to this problem is to ensure that the ground wire is of low impedance at all operating frequencies by using proper grounding techniques, including shorter ground leads and bonding equipment together.

What is a possible effect of a resonant ground connection? (G4C06)

A. Overheating of ground straps
B. Corrosion of the ground rod
C. High RF voltages on the enclosures of station equipment
D. A ground loop

A resonant ground connection can also create problems by introducing high RF voltages onto the enclosures of station equipment. Resonance occurs when the length of the ground wire or system matches the wavelength of the operating frequency, creating a situation where RF energy is trapped or magnified in the system. This can lead to high RF voltages that are not safely dissipated, which increases the risk of RF burns and interference. To avoid this, ground connections should be made so they do not resonate with operating frequencies, ensuring safer and more stable station performance.

Lightening Ground Setup

A lightning ground setup is a system designed to safely direct lightning strikes into the ground. It prevents electrical surges from damaging equipment or posing a safety risk. This system is essential for protecting structures, antennas, and radio stations from lightning's destructive power.

Components of a Lightning Ground Setup:

1. **Lightning Rod or Air Terminal:** This is a metal rod installed at the highest point of a structure (e.g., the top of a radio tower or building). The purpose is to attract lightning strikes and provide a path for the energy to follow.

2. **Down Conductors:** Once the rod captures lightning, the down conductors carry the electrical energy from the rod to the ground. These conductors are typically made of copper or aluminum and are securely attached to the structure, ensuring a low-resistance path for the electricity to travel.

3. **Ground Rods (or Grounding Electrodes):** Ground rods direct energy into the ground. These rods are buried deep in the earth, providing a place for the electricity to dissipate safely. The rods are connected to the down conductor with strong mechanical clamps (not soldered connections) to ensure the joint doesn't fail under extreme heat conditions caused by lightning strikes.

4. **Bonding:** In the setup, all metal parts of the building and the station's equipment should be bonded or electrically connected to ensure that there are no potential differences between them. This prevents dangerous surges and reduces the chance of lightning damaging equipment.

A proper lightning grounding setup ensures that the electrical energy from a lightning strike is safely diverted into the earth, reducing the risk of fire, equipment damage, or harm to individuals in the area.

This setup is particularly critical for ham radio operators with tall antennas or towers, as these are prime targets for lightning strikes. Proper grounding protects equipment and ensures compliance with safety standards and local regulations.

Why should soldered joints not be used in lightning protection ground connections? (G4C07)

A. A soldered joint will likely be destroyed by the heat of a lightning strike
B. Solder flux will prevent a low conductivity connection
C. Solder has too high a dielectric constant to provide adequate lightning protection
D. All these choices are correct

When setting up a lightning protection system for a radio station, it is crucial to avoid soldered joints in ground connections. Soldered joints, while useful for electrical connections in many situations, are not suitable for lightning protection because of the intense heat generated by a lightning strike. The high temperature can melt or vaporize the solder, causing the joint to fail and compromising the entire grounding system. Instead, mechanical connections like clamps or compression fittings should be used, as they can withstand extreme conditions and remain intact during a lightning event.

Ground Fault Circuit Interrupter

A Ground Fault Circuit Interrupter (GFCI) is a critical safety device designed to protect individuals from electric shock. It monitors the flow of electricity through a circuit, specifically ensuring that the current flowing in the hot and neutral wires is balanced. If an imbalance occurs, such as when current flows from the hot wire to the ground, the GFCI immediately disconnects power to prevent injury or damage.

Which of the following conditions will cause a ground fault circuit interrupter (GFCI) to disconnect AC power? (G0B05)

A. Current flowing from one or more of the hot wires to the neutral wire
B. Current flowing from one or more of the hot wires directly to ground
C. Overvoltage on the hot wires
D. All these choices are correct

When the GFCI detects some electrical current flowing through an unintended path, such as a person or direct ground contact, it quickly interrupts the circuit, cutting off the electricity supply to prevent potential electric shock or harm. This is a vital safety feature for areas where electricity may come into contact with water, such as kitchens, bathrooms, and outdoor outlets.

Bonding

Bonding is the process of electrically connecting all metal components together to ensure they have the same electrical potential. This helps prevent differences in voltage between metal parts, which could otherwise result in sparks or dangerous electrical discharge. Bonding ensures that all pieces of equipment are at the same potential, reducing the risk of shock or interference.

Bond all equipment enclosures together to minimize RF "hot spots" in an amateur radio station. RF "hot spots" occur when uneven distributions of RF energy create areas of high radio frequency energy within the station, which can lead to interference or even electric shock. By bonding all equipment enclosures together, you ensure that all metal surfaces within the station are electrically connected. This helps to equalize the potential across all equipment, eliminating areas where stray RF energy could accumulate.

What technique helps to minimize RF "hot spots" in an amateur station? (G4C11)

A. Building all equipment in a metal enclosure
B. Using surge suppressor power outlets
C. Bonding all equipment enclosures together
D. Placing low-pass filters on all feed lines

Ground Loop

A ground loop occurs when multiple ground paths in a system create a difference in potential between different points, leading to undesirable currents flowing through the system. This typically happens when equipment in an audio or radio setup is grounded at multiple points that are at different electrical potentials. The difference in potential causes current to flow through the ground connections, creating a loop. This can introduce hum, noise, or interference in the system, particularly in audio signals or transmitted radio waves.

Ground loops can interfere with transmitted and received signals in an amateur radio station or introduce audible "hum" in audio equipment. To avoid ground loops, bond all equipment to a single, common ground point and ensure that no multiple ground paths exist at different potentials.

By addressing ground loops, operators can maintain cleaner signals and reduce interference, ensuring better station performance.

How can the effects of ground loops be minimized? (G4C09)

A. Connect all ground conductors in series
B. Connect the AC neutral conductor to the ground wire
C. Avoid using lock washers and star washers when making ground connections
D. Bond equipment enclosures together

A common technique for minimizing the effects of ground loops is bonding all the equipment enclosures together. By bonding, all the equipment is electrically connected to a single point, creating a unified ground reference. This eliminates differences in electrical potential between different devices, preventing unwanted currents from flowing between them and reducing the chance of interference. This technique ensures a clean and stable operating environment, improving both signal quality and equipment safety.

Making Repairs Safely

Lead-Tin Solder Safety

Lead-tin solder is commonly used in electronics, but mishandling it can lead to contamination. The primary danger from lead is its potential to contaminate food or surfaces, which can cause lead poisoning if ingested.

Which of the following is a danger from lead-tin solder? (G0B10)

A. Lead can contaminate food if hands are not washed carefully after handling the solder
B. High voltages can cause lead-tin solder to disintegrate suddenly
C. Tin in the solder can "cold flow," causing shorts in the circuit
D. RF energy can convert the lead into a poisonous gas

Lead-tin solder contains lead, a toxic substance that can be harmful if ingested. <u>After handling solder, it is crucial to wash your hands thoroughly before eating or touching food.</u> Failing to do so can cause lead particles to contaminate food, leading to long-term health risks like lead poisoning. Proper hygiene and cleanliness are essential to safely working with lead-based materials.

Chapter Summary

Safety is the foundation of every successful amateur radio operation, and this chapter has provided the essential knowledge and practices to ensure a secure environment for both you and your equipment. From grounding and bonding techniques to preventing RF exposure and managing high-voltage risks, these principles safeguard against accidents and maximize station efficiency. By adhering to safety standards, following the National Electrical Code, and performing regular evaluations, you can confidently explore the expanded opportunities that come with your General Class license. Prioritizing safety not only protects your station but also guarantees years of enjoyable, worry-free communication.

Chapter 19

Antenna Safety

BUILDING ON THE BASICS

The Technician Class exam introduced you to foundational antenna safety concepts, like grounding and basic lightning protection. As you advance to the General Class level, we'll take a deeper dive into these topics, expanding your knowledge to ensure you're prepared for more complex installations and operations. This chapter focuses on critical practices for antenna safety, including managing indoor transmitting setups, implementing advanced lightning protection systems, and optimizing grounding and bonding techniques. We'll also cover routine maintenance for high structures like towers, emphasizing the importance of preparation and proper equipment. By mastering these safety measures, you'll protect your station and yourself, equipping you to operate with confidence and expertise in any scenario.

Indoor Transmitting Antenna

When installing an indoor transmitting antenna, it's critical to consider safety, particularly concerning exposure to electromagnetic radiation. The Maximum Permissible Exposure (MPE) limits must not be exceeded in areas where people will be present. The FCC defines MPE limits to ensure individuals are not exposed to unsafe radio frequency (RF) energy levels. This is especially important indoors because there may be prolonged exposure to the RF fields generated by the antenna, which could be harmful if the levels are too high.

What precaution should be taken if you install an indoor transmitting antenna? (G0A11)

A. Locate the antenna close to your operating position to minimize feedline radiation
B. Position the antenna along the edge of a wall to reduce parasitic radiation
C. Make sure that MPE limits are not exceeded in occupied areas
D. Make sure the antenna is properly shielded

Factors like the power output, frequency of operation, antenna placement, and distance from occupied areas must be considered to ensure MPE limits are not exceeded. Proper placement and minimizing transmission power are key ways to stay within safe exposure limits, protecting anyone in the room or nearby.

MPE is a crucial safety guideline for all amateur radio operations, particularly in confined spaces with a greater chance of high exposure. Indoor installations of transmitting antennas should always prioritize health and safety by ensuring RF exposure is below the limits set by regulatory authorities. These limits are designed to prevent long-term damage from exposure to RF energy, which can affect body tissues, especially during prolonged periods of use.

Lightning Protection

Lightning Protection Ground System

A key aspect of station safety is protecting equipment and operators from lightning strikes. System grounding is vital to safely direct any lightning strikes into the earth and away from station equipment.

Where should the station's lightning protection ground system be located? (G0B04)

A. As close to the station equipment as possible
B. Outside the building
C. Next to the closest power pole
D. Parallel to the water supply line

The station's lightning protection ground system should always be located outside the building to ensure that any potential strike is diverted away from interior equipment and occupants. This external ground system provides a path for the high-voltage energy to flow safely into the earth, minimizing the risk of electrical damage or fire inside the structure.

Lightning Arrestor Placement

Lightning arrestors are critical for protecting your equipment from voltage spikes caused by nearby lightning strikes. Correct placement of these arrestors can differentiate between a functional system and severe damage.

Where should lightning arrestors be located? (G0B13)

A. Where the feed lines enter the building
B. On the antenna, opposite the feed point
C. In series with each ground lead
D. At the closest power pole ground electrode

Lightning arrestors should be installed where the feed lines enter the building. This strategic placement ensures that any voltage surge due to a lightning strike is intercepted before it reaches indoor equipment, protecting your transceivers, amplifiers, and other sensitive devices from potential destruction.

Bonding Ground Rods for Lightning Protection

Proper bonding of ground rods is necessary to maintain a single, unified grounding system, ensuring safety and efficiency in lightning protection.

Which of the following is required for lightning protection ground rods? (G0B11)

A. They must be bonded to all buried water and gas lines
B. Bends in ground wires must be made as close as possible to a right angle
C. Lightning grounds must be connected to all ungrounded wiring
D. They must be bonded together with all other grounds

For effective lightning protection, <u>all ground rods must be bonded with all other grounds</u>, such as electrical system grounds and the station's RF grounds. This interconnected grounding system ensures that there are no voltage differences between grounds, minimizing the risk of damage from a lightning strike by allowing the energy to flow evenly and safely into the earth.

Routine Maintenance

Maintaining your antennas and feedlines is as critical as caring for your transceivers. Start by inspecting your antennas for any physical damage or corrosion. Look for signs of wear, such as rust on metal parts or cracks in plastic components. Ensuring that all elements are correctly aligned and tensioned is also critical. Misaligned elements can affect the performance of your antenna, reducing its efficiency. Use a level and tension gauge to check and adjust the alignment and tension as needed.

Cleaning and tightening connections and insulators is another vital task. Dirty or loose connections can lead to poor signal quality and increased SWR (Standing Wave Ratio). Use a wire brush or emery cloth to clean metal connectors and apply a small amount of dielectric grease to prevent corrosion. Tighten all bolts and screws to ensure a secure connection. Checking the insulators for cracks or dirt buildup is also necessary, as these can cause electrical leakage or arcing.

Climbing a Tower

Climbing towers, such as those used for ham radio antennas, require strict attention to safety to prevent accidents. An essential part is using a properly rated safety harness designed explicitly for tower work. <u>The harness should be inspected before each use to ensure it is within its service life, free from damage, and correctly rated for the climber's weight</u>. Proper use of a safety harness helps ensure that the climber is protected from severe injury in the event of a fall.

> **Which of these choices should be observed when climbing a tower using a safety harness? (G0B07)**
>
> A. Always hold on to the tower with one hand
> **B. Confirm that the harness is rated for the weight of the climber and that it is within its allowable service life**
> C. Ensure that all heavy tools are securely fastened to the harness
> D. All these choices are correct

The harness must be able to support the climber's weight and remain in good condition. Harnesses can degrade over time, so confirming that they are within their service life ensures they will function properly if needed.

Lockout-Tagout Safety

The risk of accidental contact with electrical currents is high when climbing a tower that supports electrically powered devices. Before ascending the tower, it is crucial to follow lockout-tagout procedures, which involve <u>cutting off the electrical power supply and clearly tagging it to prevent accidental reactivation.</u> This eliminates the risk of electrocution while working on or near the tower.

What should be done before climbing a tower that supports electrically powered devices? (G0B08)

A. Notify the electric company that a person will be working on the tower
B. Make sure all circuits that supply power to the tower are locked out and tagged
C. Unground the base of the tower
D. All these choices are correct

By ensuring all power circuits are disabled and tagged, tower climbers can safely work without encountering live electrical wires or components.

Chapter Summary

To conclude, ensuring antenna safety requires a careful balance of protective measures, regular maintenance, and adherence to established standards. From managing indoor antennas and preventing RF exposure to implementing robust lightning protection systems, each aspect of antenna safety is crucial for safeguarding the operator and equipment. Routine maintenance, especially when working with towers, demands meticulous attention to protocols to avoid accidents. By following these guidelines, radio operators can maintain a safe and reliable station that minimizes risk and maximizes performance, allowing them to focus on practical and enjoyable communication.

Operating Your Radio

MASTERING THE OPERATION OF your radio is key to becoming a skilled and confident ham operator. This section covers essential practices and knowledge, from proper operating procedures and understanding ham radio jargon to advanced CW operations and digital modes. You'll also learn how to manage interference, ensuring clear and reliable communication. Whether you're refining your skills or tackling new challenges, this section equips you with practical tools and techniques to ace your exam.

Chapter 20

Ham Radio Jargon

THIS CHAPTER BUILDS ON the Q codes you learned in the Technician's exam. Those were QRM and QSY. These three-letter codes act as shorthand to convey common phrases, making communication more efficient and precise. Q codes are understood universally in amateur radio, even across different languages, and they help streamline conversations, particularly in situations where speed and clarity are important. By mastering these signals, operators can enhance their CW proficiency and improve communication flow during their transmissions.

While all the codes can be useful, just like in the tech exam prep, let's focus on the ones you will need for the General exam. You will learn the rest over time through first-hand use.

Commonly Used Q Signals

What does the Q signal "QRL?" mean? (G2C04)

A. "Will you keep the frequency clear?"
B. "Are you operating full break-in?" or "Can you operate full break-in?"
C. "Are you listening only for a specific station?"
D. "Are you busy?" or "Is this frequency in use?"

The Q signal "QRL?" asks, "Are you busy?" or "Is this frequency in use?". It is a common courtesy for operators to send "QRL?" before starting a transmission to ensure they are not interfering with another ongoing communication. This helps prevent unintentional disruptions in radio communications, especially in busy bands.

What does the Q signal "QRV" mean? (G2C11)

A. You are sending too fast
B. There is interference on the frequency
C. I am quitting for the day
D. I am ready to receive

The Q signal "QRV" means "I am ready to receive." This signal is typically sent to indicate readiness to start receiving information from another station. It lets the operator know the receiver is standing by and prepared to copy the transmitted message.

What does the Q signal "QSL" mean? (G2C09)

A. Send slower
B. We have already confirmed the contact
C. I have received and understood
D. We have worked before

The Q signal "QSL" means "I have received and understood." This is an acknowledgment that the message has been successfully received. It's often used in live communication and the exchange of physical QSL cards to confirm contact between stations.

Troubleshooting and Adjustments Q Codes

What does the Q signal "QRN" mean? (G2C10)

A. Send more slowly
B. Stop sending
C. Zero beat my signal
D. I am troubled by static

The Q signal "QRN" means "I am troubled by static." This signal reports interference or noise in the receiver's environment, which may make it difficult to copy the signal clearly. Informing the other station of QRN can lead to adjusting transmission methods to overcome static or noise.

What should you do if a CW station sends "QRS?" (G2C02)

A. Send slower
B. Change frequency
C. Increase your power
D. Repeat everything twice

The Q signal "QRS?" is a request to "Send slower". This is used when the receiving station is having difficulty copying the transmission due to the speed of the code. It helps maintain clear and accurate communication by ensuring the message is understood.

Understanding the Q-codes helps with smooth and efficient communication. But you only need to memorize a few for the exam. Memorize those now, and the rest will come with use.

QRP Operation

What is QRP operation? (G2D10)

A. Remote piloted model control
B. Low-power transmit operation
C. Transmission using Quick Response Protocol
D. Traffic relay procedure net operation

QRP operation refers to low-power transmit operation in amateur radio. The term "QRP" comes from the Q code system, which originally meant "reduce power." QRP operators typically transmit with 5 watts or less for CW and digital modes or 10 watts or less for SSB. QRP aims to make efficient contacts using minimal power, challenging operators to improve their operating skills, antenna systems, and station efficiency.

This mode of operation is favored by many amateur radio operators who enjoy the challenge of making successful long-distance contacts with minimal power output. QRP enthusiasts often take pride in their ability to communicate over great distances with very little power, which requires skillful operating techniques, good propagation, and efficient antennas. The QRP community is a vibrant part of amateur radio, emphasizing efficiency and the thrill of low-power communication.

In contests and regular QSOs, QRP operators often note their low-power status, which adds to the challenge and sense of accomplishment when a contact is made under those conditions.

NATO Phonetic Alphabet

The NATO Phonetic Alphabet is an internationally recognized set of words used to clearly communicate letters over radio, telephone, or other audio transmissions. It helps reduce confusion, especially in noisy environments or when signals are weak. Each letter of the alphabet is represented by a distinct word, ensuring that the intended letters are still clear even if the transmission quality is poor. The military, aviation, emergency services, and amateur radio operators worldwide use the phonetic alphabet to enhance communication clarity.

In the NATO Phonetic Alphabet, each letter is assigned a word that sounds distinctly different from the others. For example, "A" is Alpha, "B" is Bravo, "C" is Charlie, and "D" is Delta. These specific words were chosen to minimize confusion and misunderstanding, especially in critical situations where clarity is paramount. This system allows radio operators to spell out words and call signs clearly, ensuring the message is accurately received regardless of signal interference or noise.

> **Which of the following are examples of the NATO Phonetic Alphabet? (G2D07)**
>
> A. Able, Baker, Charlie, Dog
> B. Adam, Boy, Charles, David
> C. America, Boston, Canada, Denmark
> **D. Alpha, Bravo, Charlie, Delta**

ALPHA—that's the key to this question. Remember that 'A' is alpha, and you have mastered this question.

Resources

Chapter Summary

In amateur radio, Q-codes and the NATO phonetic alphabet serve as valuable tools for clear and efficient communication. Q-codes simplify complex messages into easy-to-understand shorthand, while the NATO alphabet ensures accuracy in conveying critical information like call signs and locations. Mastering these tools enhances your operating skills and ensures you're understood in any situation, fostering better communication with operators worldwide.

Chapter 21

Voice Communication

Station Logs

Keeping a station log is a common and recommended practice among amateur radio operators. A station log is a record of contacts made, equipment used, and important events related to the station's operation. While not always required by FCC rules, maintaining a log helps ensure that operators have a detailed account of their radio activities, which can be beneficial in a variety of circumstances, especially if questions arise about their transmissions or station operations.

Why do many amateurs keep a station log? (G2D08)

A. The FCC requires a log of all international contacts
B. The FCC requires a log of all international third-party traffic
C. The log provides evidence of operation needed to renew a license without retest
D. To help with a reply if the FCC requests information about your station

---

Many amateur radio operators keep a station log to have a detailed record of their communications. This can be especially useful if the FCC requests information about your station's activities. Suppose there's a need to verify that a transmission was made correctly. In that case, the log provides important details such as the frequency, mode, and time of communication. Additionally, keeping a log helps with personal record-keeping, tracking DX contacts, and organizing information for awards programs. A comprehensive log can also protect you in case of any dispute regarding interference or using specific frequencies.

Voice

When choosing a frequency to initiate a call, it is important to follow the voluntary band plan so much so that there is an exam question about it! Band plans are community-agreed guidelines that organize the amateur radio spectrum into specific sections based on usage, such as SSB, CW, or digital modes. By adhering to the band plan, operators help avoid interference with ongoing communications and ensure that they are operating in a manner that is respectful to others using the same spectrum. For example, calling a CQ on a frequency dedicated to digital modes with a voice transmission would cause unnecessary interference and go against commonly accepted practices.

Which of the following complies with commonly accepted amateur practice when choosing a frequency on which to initiate a call? (G2B07)

A. Listen on the frequency for at least two minutes to be sure it is clear
B. Identify your station by transmitting your call sign at least 3 times
C. Follow the voluntary band plan
D. All these choices are correct

Following the voluntary band plan also ensures that operators respect frequencies designated for specific purposes, like emergency communication or contesting. These guidelines help maintain an orderly use of the limited radio spectrum and make it easier for everyone to find and use the correct frequencies for their communication needs.

So, based on that info.

What is the voluntary band plan restriction for US stations transmitting within the 48 contiguous states in the 50.1 MHz to 50.125 MHz band segment? (G2B08)

A. Only contacts with stations not within the 48 contiguous states
B. Only contacts with other stations within the 48 contiguous states
C. Only digital contacts
D. Only SSTV contacts

The 50 MHz band, often referred to as the 6-meter band, is a unique section of the amateur radio spectrum that offers opportunities for long-distance communication, especially during periods of sporadic E propagation and solar activity. While the band is open to various types of communication, certain segments within it, such as the 50.1 MHz to 50.125 MHz range, have specific restrictions. These rules help ensure orderly and effective communication, particularly for DX (long-distance) contacts.

In the 50.1 MHz to 50.125 MHz band segment, US stations within the 48 contiguous states are limited to making contact only with stations outside the 48 contiguous states. This voluntary band plan restriction is designed to reserve this band portion for long-distance or DX communications, encouraging contacts with stations in Alaska, Hawaii, or other countries. By reserving this segment for DX contacts, operators avoid crowding the band with local communications, thus maximizing the chances for making rare, long-distance connections.

This restriction also helps reduce interference in a portion of the 6-meter band that is highly sought after for DX activity, especially during favorable propagation conditions. Remembering this for the exam is key, as it highlights how band segments are managed to ensure fair use and promote long-distance communication on specific frequencies.

Access to Frequencies

All operators share the same frequency spectrum, and there are no reserved or private frequencies for individual operators or groups. The principle of frequency sharing is fundamental to amateur radio, meaning that frequencies are generally available on a first-come, first-served basis. The only exception to this rule is during emergencies, where communication for emergency services takes precedence over regular operations.

When it comes to accessing frequencies, it is important to remember that, except during emergencies, no amateur station has priority access to any frequency. Any operator may use an available frequency, provided it is not currently in use. In the event that two stations wish to use the same frequency, it is customary to share or find another available frequency. However, if a frequency is used for an emergency or disaster-related communication, those transmissions take priority over all others.

Which of the following is true concerning access to frequencies? (G2B01)

A. Nets have priority
B. QSOs in progress have priority
C. Except during emergencies, no amateur station has priority access to any frequency
D. Contest operations should yield to non-contest use of frequencies

This principle ensures fair access for all operators and promotes cooperation within the amateur radio community. It is also an important guideline to recall for the exam, as understanding the rules for frequency access helps maintain smooth operations on the airwaves and prevents disputes between operators.

Receiver Bandwidth

Receiver bandwidth refers to the range of frequencies that a receiver can process at one time. Different operating modes like CW, SSB, and FM have unique bandwidth requirements. It's important to match the receiver's bandwidth to the operating mode's bandwidth for optimal performance. This ensures that the receiver only processes the necessary portion of the radio spectrum, helping to minimize noise and interference and allowing for clearer reception of the intended signal.

Matching the receiver bandwidth to the bandwidth of the operating mode is important because it results in the best sig-nal-to-noise ratio (SNR). The signal-to-noise ratio measures how much the desired signal is received compared to background noise. If the receiver's bandwidth is too wide, it picks up unnecessary noise and interference from frequencies outside of the desired signal range, reducing the clarity of the transmission. By narrowing the bandwidth to just the amount needed for the mode you're using, you filter out unwanted noise, improving the clarity and overall quality of the received signal.

Why is it good to match receiver bandwidth to the bandwidth of the operating mode? (G8B09)

A. It is required by FCC rules
B. It minimizes power consumption in the receiver
C. It improves impedance matching of the antenna
D. It results in the best signal-to-noise ratio

To remember this for the exam, think of matching the bandwidth like "zooming in" on just the right portion of the signal, reducing noise, and making the signal stand out more clearly. This technique helps ensure the best listening experience with minimal interference.

Frequency Etiquette

Proper frequency etiquette is crucial to being a responsible amateur radio operator. Before transmitting on a frequency, operators should always check if the frequency is clear to avoid harmful interference with ongoing communications. Additionally, all amateur radio operators are expected to follow specific emergency protocols, such as prioritizing stations in distress. Furthermore, good amateur practice involves handling unexpected interference during communications and effectively managing nets to ensure smooth operation, even in poor conditions.

Let's review some radio etiquette that you will see on the exam.

How can you avoid harmful interference on an apparently clear frequency before calling CQ on CW or phone? (G2B06)

A. Send "QRL?" on CW, followed by your call sign; or, if using phone, ask if the frequency is in use, followed by your call sign
B. Listen for 2 minutes before calling CQ
C. Send the letter "V" in Morse code several times and listen for a response, or say "test" several times and listen for a response
D. Send "QSY" on CW or if using phone, announce "the frequency is in use," then give your call sign and listen for a response

To ensure that you are not interfering with ongoing communications, it is good practice to first check if the frequency is in use. For CW (Morse code), send the code "QRL?" followed by your call sign. On phone (voice), simply ask, "Is the frequency in use?" followed by your call sign. This helps prevent accidental interruptions of communications that may not be immediately audible.

What is the first thing you should do if you are communicating with another amateur station and hear a station in distress break in? (G2B02)

A. Inform your local emergency coordinator
B. Acknowledge the station in distress and determine what assistance may be needed
C. Immediately decrease power to avoid interfering with the station in distress
D. Immediately cease all transmissions

What is the first thing you should do if you are communicating with another amateur station and hear a station in distress break in? If a station in distress breaks in during your communication, the first and most important step is to acknowledge the station in distress. You should then determine what assistance may be required. In emergency situations, the distress signal takes priority, and all other communication should cease to allow assistance.

What is good amateur practice if propagation changes during a contact creating interference from other stations using the frequency? (G2B03)

A. Advise the interfering stations that you are on the frequency and that you have priority
B. Decrease power and continue to transmit
C. Attempt to resolve the interference problem with the other stations in a mutually acceptable manner
D. Switch to the opposite sideband

If propagation conditions change during a contact, leading to interference from other stations, it is good practice to attempt to resolve the interference issue in a mutually acceptable manner. This may involve moving to a new frequency or adjusting your equipment settings to minimize interference. Cooperation is key to ensuring all operators can continue communicating effectively.

Which of the following is good amateur practice for net management? (G2B10)

A. Always use multiple sets of phonetics during check-in
B. Have a backup frequency in case of interference or poor conditions
C. Transmit the full net roster at the beginning of every session
D. All these choices are correct

A key aspect of effective net management is having a backup frequency in case of interference or poor conditions. This ensures that the net can continue without interruption if the main frequency becomes unusable. Planning a backup frequency ensures successful and efficient net operation.

These principles and always prioritizing emergency communication help maintain the integrity and usefulness of the amateur radio spectrum for everyone.

Breaking Into a Phone Contact

It's common for operators to want to join or break into an ongoing conversation, also known as a QSO. It's important to follow established etiquette to avoid disrupting the conversation or causing interference when trying to do so. In phone (voice) contacts, there is a simple and polite method for breaking into a conversation, allowing you to join the QSO while maintaining good operating practices.

The recommended way to break into a phone contact is to say your call sign once. This brief and straightforward method alerts the operators in the ongoing conversation that someone wishes to join without interrupting their communication flow. By simply stating your call sign, you respect the conversation and give others the chance to invite you to join when they're ready.

What is the recommended way to break into a phone contact? (G2A08)

A. Say "QRZ" several times, followed by your call sign
B. Say your call sign once
C. Say "Breaker Breaker"
D. Say "CQ" followed by the call sign of either station

This practice ensures that you follow proper on-air etiquette and allows for smooth and courteous transitions when operators wish to join a QSO. Avoid saying "break" or other words that may be confusing or disruptive. Simply state your call sign clearly and wait for an acknowledgment.

And I appreciate the exam creator's sense of humor in including "breaker breaker!"

Another exam question similar to this:

Which of the following indicates that you are looking for an HF contact with any station? (G2D05)

A. Sign your call sign once, followed by the words "listening for a call" -- if no answer, change frequency and repeat
B. Say "QTC" followed by "this is" and your call sign -- if no answer, change frequency and repeat
C. Repeat "CQ" a few times, followed by "this is," then your call sign a few times, then pause to listen, repeat as necessary
D. Transmit an unmodulated carried for approximately 10 seconds, followed by "this is" and your call sign, and pause to listen -- repeat as necessary

---

Now, don't let the HF part of this question fool you. When you're open to contact with any station on any band, you send out a CQ call. "CQ" is a general call for communication, meaning you're looking for anyone to respond. This call is standard practice across the ham radio community and is crucial for starting conversations with fellow operators worldwide. It's an efficient way to reach out, especially when you're not looking for any particular station or operator.

To make a CQ call, the operator should follow a structured format: Repeat "CQ" a few times, followed by "this is," and then your call sign a few times. After sending the call, pause for a moment to listen for any responses. If no one replies, you can repeat the process until you establish contact with another operator. This method ensures your call is heard clearly and provides ample opportunity for others to respond. It's also important to ensure that the frequency is clear before making your call by asking, "Is this frequency in use?" to avoid interference.

DX Calls

In amateur radio, operators often seek to establish DX contacts, which means communicating with stations in distant locations, typically outside their own country or region. When an operator calls CQ DX, they are specifically requesting responses from stations that are far away, particularly outside of their immediate geographical area. Operators in the contiguous 48 states (the lower 48 in the US) are looking to contact stations outside this region.

When a station in the contiguous 48 states calls "CQ DX," only stations outside the lower 48 should respond. This is because the calling station is specifically seeking long-distance or international contacts, and responding from within the lower 48 would not fulfill the DX objective. Stations from Alaska, Hawaii, US territories, and other countries are typically the intended responders.

> **Generally, who should respond to a station in the contiguous 48 states calling "CQ DX"? (G2A11)**
>
> A. Any caller is welcome to respond
> B. Only stations in Germany
> **C. Any stations outside the lower 48 states**
> D. Only contest stations

If you're in the lower 48, you should wait for another opportunity to call unless you are located outside of that region. This practice respects the operator's goal of seeking DX communication and follows proper on-air etiquette.

Single Sideband

Single Sideband (SSB) is a widely used voice communication mode in HF amateur bands. It is a variation of amplitude modulation (AM) that offers several advantages, including greater power efficiency and reduced bandwidth usage. Instead of transmitting both sidebands and the carrier as in traditional AM, SSB transmits only one sideband (upper or lower). In contrast, the other sideband and carrier are suppressed. This makes SSB an ideal mode for long-distance communication on crowded HF bands.

The most commonly used voice mode in HF amateur bands is single-sided band (SSB). SSB is preferred due to its power and bandwidth efficiency, making it ideal for long-distance contacts over the HF spectrum. By transmitting only one sideband, SSB allows for clearer communication with less interference.

Which mode of voice communication is most commonly used on the HF amateur bands? (G2A05)

A. Frequency modulation
B. Double sideband
C. Single sideband
D. Single phase modulation

The key characteristic of Single Sideband is that only one sideband is transmitted. In contrast, the other sideband and the carrier are suppressed. This selective transmission reduces the bandwidth used and focuses the power on a single sideband, allowing for more efficient use of the radio spectrum and minimizing interference with other signals.

Which of the following statements is true of single sideband (SSB)? (G2A07)

A. Only one sideband and the carrier are transmitted; the other sideband is suppressed
B. Only one sideband is transmitted; the other sideband and carrier are suppressed
C. SSB is the only voice mode authorized on the 20-, 15-, and 10-meter amateur bands
D. SSB is the only voice mode authorized on the 160-, 75-, and 40-meter amateur bands

One of the main advantages of SSB over other analog voice modes like AM or FM is that it uses less bandwidth and provides greater power efficiency. This means more of the transmitter's power is dedicated to the signal being sent, resulting in clearer, stronger signals over longer distances. Additionally, the reduced bandwidth allows more signals to fit into a given portion of the HF spectrum.

Which of the following is an advantage of using single sideband, as compared to other analog voice modes on the HF amateur bands? (G2A06)

A. Very high-fidelity voice modulation
B. Less subject to interference from atmospheric static crashes
C. Ease of tuning on receive and immunity to impulse noise
D. Less bandwidth used and greater power efficiency

SSB is the dominant mode for voice communication on HF amateur bands because it combines bandwidth conservation with power efficiency. Operators can communicate more effectively by suppressing the carrier and one sideband, especially in long-distance and crowded band conditions.

Automatic Level Control

In single sideband (SSB) transmissions, Automatic Level Control (ALC) ensures that the transmitter operates within safe power levels without causing distortion. ALC helps regulate the audio signal input to prevent overdriving the transmitter, which could result in signal distortion or interference. Adjusting the ALC properly helps maintain clear and clean transmissions, optimizing the performance of your SSB transceiver.

The transmit audio or microphone gain control is typically adjusted to properly set the ALC on a single sideband transceiver. This adjustment controls the audio level fed into the transmitter. The operator can ensure that the signal remains within the proper ALC range by increasing or decreasing the microphone gain. Too high a microphone gain can push the ALC beyond its limits, causing distortion, while too low a gain may result in weak transmissions.

What control is typically adjusted for proper ALC setting on a single sideband transceiver? (G2A12)

A. RF clipping level
B. Transmit audio or microphone gain
C. Antenna inductance or capacitance
D. Attenuator level

For optimal performance, the ALC should be adjusted to reach the recommended level during normal speech peaks, ensuring clear transmission and avoiding unnecessary over-modulation or distortion. This makes the transmit audio or microphone gain the key control to adjust for achieving the correct ALC settings.

Lower Sideband

Lower Sideband (LSB) is a type of Single Sideband (SSB) modulation that is widely used in voice communications on certain HF bands. In amateur radio, specific sidebands are typically used on different bands to standardize operations and avoid interference. For the 160-meter, 75-meter, and 40-meter bands, the lower sideband (LSB) is the preferred mode of voice communication. This practice ensures consistency and better signal management in the crowded HF bands, particularly for amateur operators.

The most commonly used mode for voice communication on the 160-, 75-, and 40-meter bands is Lower Sideband (LSB). In LSB transmission, only the lower sideband of the signal is transmitted, while the carrier and upper sideband are suppressed. This saves bandwidth and improves power efficiency, allowing for clearer communication, particularly over long distances.

Which mode is most commonly used for voice communications on the 160-, 75-, and 40-meter bands? (G2A02)

A. Upper sideband
B. Lower sideband
C. Suppressed sideband
D. Double sideband

And why is that?

Most amateur stations use Lower Sideband (LSB) on the 160-, 75-, and 40-meter bands because it is a commonly accepted amateur practice. This convention has been established to maintain consistency and order on the airwaves. By adhering to this practice, operators reduce confusion and interference when tuning into different bands, as they know what mode to expect for voice communications.

Why do most amateur stations use lower sideband on the 160-, 75-, and 40-meter bands? (G2A09)

A. Lower sideband is more efficient than upper sideband at these frequencies
B. Lower sideband is the only sideband legal on these frequency bands
C. Because it is fully compatible with an AM detector
D. It is commonly accepted amateur practice

Historically, Lower Sideband (LSB) was adopted for frequencies below 9 MHz, including the 160-, 75-, and 40-meter bands. Upper Sideband (USB) became the norm for frequencies above 9 MHz. This separation in sideband usage helps maintain efficiency and minimizes conflicts between operators on adjacent frequencies, making it easier to communicate clearly.

Lower Sideband (LSB) use on the 160-, 75-, and 40-meter bands is part of a long-standing tradition in amateur radio. By standardizing the use of sidebands, the amateur radio community has created a more predictable operating environment. This division between LSB on lower frequency bands and USB on higher frequency bands is part of accepted best practices, making it easier for operators to communicate effectively without unnecessary interference.

Now, some questions.

How close to the lower edge of a band's phone segment should your displayed carrier frequency be when using 3 kHz wide LSB? (G4D10)

A. At least 3 kHz above the edge of the segment
B. At least 3 kHz below the edge of the segment
C. At least 1 kHz below the edge of the segment
D. At least 1 kHz above the edge of the segment

When using LSB, especially near the lower edge of a band's phone segment, it's important to position your carrier frequency carefully. Since an LSB signal occupies frequencies below the carrier, your carrier frequency must be at least 3 kHz above the edge of the segment to ensure your signal doesn't extend outside the authorized band. For example, if the lower edge of the segment is 7.170 MHz, your carrier frequency should be no lower than 7.173 MHz when using a 3 kHz wide signal. This practice helps avoid transmitting outside the allocated band and ensures compliance with FCC regulations.

What frequency range is occupied by a 3 kHz LSB signal when the displayed carrier frequency is set to 7.178 MHz? (G4D08)

A. 7.178 MHz to 7.181 MHz
B. 7.178 MHz to 7.184 MHz
C. 7.175 MHz to 7.178 MHz
D. 7.1765 MHz to 7.1795 MHz

When you're transmitting on Lower Sideband (LSB), the signal occupies a range of frequencies below the displayed carrier frequency. For example, if your radio's carrier frequency is set to 7.178 MHz, a 3 kHz LSB signal will occupy a frequency range from 7.175 MHz to 7.178 MHz. This is because the LSB signal extends downward from the carrier frequency by 3 kHz. Understanding this relationship between your displayed frequency and the actual bandwidth occupied is critical for ensuring you do not interfere with signals on neighboring frequencies, especially near the band edges.

Both of these questions highlight the importance of understanding how LSB signals use bandwidth and how to properly position your carrier frequency on the band. To avoid interference or rule violations, you must maintain awareness of your signal's actual footprint, especially when operating near the band's edges.

Upper Sideband

In amateur radio, Upper Sideband (USB) is the most commonly used mode for Single Sideband (SSB) voice communication on VHF and UHF bands, as well as frequencies 14 MHz and higher. This convention arose to standardize operations, prevent interference, and maintain clear communication across the spectrum. By using USB on higher frequencies, operators ensure that their transmissions are compatible with standard practices, making communication easier and more reliable on these bands.

The standard mode for SSB voice communication on VHF and UHF bands is Upper Sideband (USB). USB is preferred for frequencies in these bands due to historical and technical reasons, including better signal clarity and ease of operation performance.

Which mode is most commonly used for SSB voice communications in the VHF and UHF bands? (G2A03)

A. Upper sideband
B. Lower sideband
C. Suppressed sideband
D. Double sideband

For voice communications on frequencies 14 MHz and higher, such as the 20-meter band, Upper Sideband is the most commonly used mode. This practice ensures uniformity and prevents confusion, as operators know that USB is the expected mode on these higher frequency bands.

Which mode is most commonly used for voice communications on frequencies of 14 MHz or higher? (G2A01)

A. Upper sideband
B. Lower sideband
C. Suppressed sideband
D. Double sideband

Upper Sideband is also the most common mode for voice communication on bands like 17 meters and 12 meters. This is part of the broader convention in amateur radio that USB is used on bands above 9 MHz, providing consistency across various frequencies.

Which mode is most commonly used for voice communications on the 17- and 12-meter bands? (G2A04)

A. Upper sideband
B. Lower sideband
C. Suppressed sideband
D. Double sideband

In amateur radio, the choice of sideband mode for voice communications is dictated by long-standing conventions. On frequencies 14 MHz and higher, including VHF and UHF bands, USB is the standard for SSB voice communication. This ensures that operators are transmitting in a predictable manner, reducing interference and improving the quality of communications. USB is particularly advantageous for these frequencies due to its more efficient use of bandwidth and clearer signal quality, especially for crowded bands.

By following the convention of using Upper Sideband on bands like 17 meters, 12 meters, and higher frequencies, operators maintain compatibility with the broader amateur radio community, ensuring that their transmissions are understood and that communication can happen smoothly worldwide.

Now, some questions.

How close to the upper edge of a band's phone segment should your displayed carrier frequency be when using 3 kHz wide USB? (G4D11)

A. At least 3 kHz above the edge of the band
B. At least 3 kHz below the edge of the band
C. At least 1 kHz above the edge of the segment
D. At least 1 kHz below the edge of the segment

When transmitting in USB, the signal extends upward from the carrier frequency. For a 3 kHz wide USB signal, the carrier frequency must be at least 3 kHz below the upper edge of the phone segment to prevent transmitting outside the authorized band. For example, if the upper edge of the band is 14.350 MHz, your carrier frequency should be no higher than 14.347 MHz. This ensures that the 3 kHz signal doesn't exceed the band limits, which helps you comply with regulations and avoid interfering with stations on adjacent frequencies.

What frequency range is occupied by a 3 kHz USB signal with the displayed carrier frequency set to 14.347 MHz? (G4D09)

A. 14.347 MHz to 14.647 MHz
B. 14.347 MHz to 14.350 MHz
C. 14.344 MHz to 14.347 MHz
D. 14.3455 MHz to 14.3485 MHz

A 3 kHz USB signal occupies a frequency range starting from the displayed carrier frequency and extending 3 kHz upward. For instance, if your radio is set to 14.347 MHz, the actual signal will occupy the range from 14.347 MHz to 14.350 MHz. It's important to understand how USB signals extend above the carrier frequency to avoid operating too close to the band edge and accidentally transmitting outside the allowed frequency range.

Both concepts relate to the importance of properly positioning your carrier frequency when using USB. Understanding the bandwidth of your signal ensures you stay within the band limits and avoid interfering with other stations.

Bandwidth

When selecting a frequency for SSB transmissions, it is important to maintain a separation of 2 kHz to 3 kHz from other stations. SSB signals typically use bandwidths of around 2.5 kHz, so maintaining this distance helps avoid interference. If stations are spaced too closely, signals can bleed into adjacent frequencies, causing distortion or difficulty hearing both transmissions.

When selecting an SSB transmitting frequency, what minimum separation should be used to minimize interference to stations on adjacent frequencies? (G2B05)

A. 5 Hz to 50 Hz
B. 150 Hz to 500 Hz
C. 2 kHz to 3 kHz
D. Approximately 6 kHz

Following this separation guideline, operators can transmit clean, undisturbed signals while minimizing the risk of interference for other stations. This is especially important in busy bands where multiple operators are transmitting simultaneously. Ensuring adequate separation is key to maintaining orderly and efficient communication on the airwaves.

Modulation and Sideband Selection

In radio communications, modulation allows a carrier signal to carry information by modifying its properties. Balanced modulators are integral to this process, especially when creating signals used in single-sideband transmission. When a signal is modulated, multiple "sidebands" are created, and selecting the appropriate one is crucial for efficient communication. This section explores how modulation circuits work and how sidebands are selected and extracted for effective transmission and reception.

> **What circuit is used to select one of the sidebands from a balanced modulator? (G7C01)**
>
> A. Carrier oscillator
> **B. Filter**
> C. IF amplifier
> D. RF amplifier

A filter selects one of the sidebands from a balanced modulator. Balanced modulation generates both upper and lower sidebands around the carrier frequency. However, in single-sideband (SSB) transmissions, it's crucial to filter out one of these sidebands to conserve bandwidth and power. The filter allows only the desired sideband, either the upper or lower, to pass through and block the other sideband, ensuring cleaner transmission and reception.

> **What output is produced by a balanced modulator? (G7C02)**
>
> A. Frequency modulated RF
> B. Audio with equalized frequency response
> C. Audio extracted from the modulation signal
> **D. Double-sideband modulated RF**

A balanced modulator creates a double-sideband (DSB) signal, meaning it produces two identical sidebands on either side of the carrier frequency. Importantly, the carrier itself is suppressed in a balanced modulator, allowing only the modulated sidebands to be transmitted. This makes it an efficient way to prepare signals for transmission, although further filtering (to extract just one sideband) is required for single-sideband operation.

> **How is a product detector used? (G7C04)**
>
> A. Used in test gear to detect spurious mixing products
> B. Used in transmitter to perform frequency multiplication
> C. Used in an FM receiver to filter out unwanted sidebands
> **D. Used in a single sideband receiver to extract the modulated signal**

A product detector is a specialized circuit used in single-sideband receivers to demodulate signals. It works by mixing the incoming SSB signal with a locally generated signal to retrieve the original modulated information. This process allows the receiver to extract the voice or data information from the transmitted RF signal, making it essential for SSB and other types of amplitude-modulated communication systems.

Understanding how modulation circuits function, particularly in relation to sideband selection and signal extraction, is crucial for optimizing the efficiency of radio communications. Whether using filters to select the correct sideband or employing a product detector to recover transmitted data, each component ensures signals are transmitted clearly and efficiently.

Chapter Summary

This chapter has taken you through the critical aspects of voice communication in amateur radio, from selecting appropriate frequencies and understanding modulation to mastering sideband conventions and practicing proper on-air etiquette. Voice communication lies at the heart of the amateur radio experience, fostering global connections and ensuring clear, efficient exchanges. By adhering to band plans, respecting frequency etiquette, and understanding technical details like bandwidth and sideband usage, you not only improve your communication skills but also contribute to the smooth operation of HF bands. With these tools, you are well-equipped to navigate the airwaves confidently, ensuring your transmissions are effective, compliant, and a pleasure for others to hear.

Chapter 22

Continuous Wave

MORSE CODE, OR CONTINUOUS Wave (CW), remains one of amateur radio's most iconic and effective communication modes. Its simplicity, reliability, and ability to cut through noise make it a powerful tool for ham radio operators, especially in challenging conditions. For General Class operators, mastering CW goes beyond just learning the code—it's about understanding the techniques, protocols, and tools that make CW operation seamless and efficient.

In this chapter, we'll delve into key aspects of CW operation, from choosing the right frequency to understanding advanced practices like full break-in (QSK) operation and zero beating. You'll learn the meaning of common CW terms, the use of prosigns, and how to adjust your sending speed to match other operators. We'll also cover the role of equipment like electronic keyers and explore techniques to enhance signal clarity. Whether you're a seasoned CW operator or just beginning to explore its possibilities, this chapter will equip you with the knowledge and skills to confidently incorporate CW into your ham radio experience.

Basic Operations

Zero beat refers to the process of matching the transmit frequency to the frequency of a received signal. When two frequencies align, the tone becomes zero or minimal, ensuring both stations transmit on the same frequency, optimizing signal clarity.

What does the term "zero beat" mean in CW operation? (G2C06)

A. Matching the speed of the transmitting station
B. Operating split to avoid interference on frequency
C. Sending without error
D. Matching the transmit frequency to the frequency of a received signal

---

Zero beat is the process of tuning your receiver to match the frequency of a CW signal, allowing you to hear the clearest and most precise tone. This ensures your transmission is aligned with the station you're communicating with, improving signal clarity and minimizing interference.

As you tune into a CW signal, the pitch changes depending on how close you are to the correct frequency. The goal is to adjust your tuning until the tone matches a standard pitch, typically around 600-800 Hz, which corresponds to the beat frequency between your receiver's local oscillator and the incoming signal. Achieving zero beat is vital for clear communication, avoiding interference, and efficiently operating in crowded band conditions, such as during contests or DXing.

To achieve zero beat, you can manually adjust your receiver's tuning while listening for the tone to match, or use visual aids like bar graphs or indicators on modern transceivers. For transmitters without automatic offset for CW tones, aligning your transmitted signal with the receiving station's tone is necessary. Mastering zero beat is a fundamental skill in CW operation that enhances communication quality and demonstrates good operating practices.

Key Operating Practices

One key aspect of good amateur practice when operating Continuous Wave mode is selecting a transmitting frequency that minimizes interference with other stations. Adequate separation from adjacent frequencies is essential to avoid unwanted interference. Following the recommended spacing between CW signals, operators can maintain clear and effective communication without disrupting nearby stations.

When selecting a frequency to transmit CW, it is important to maintain a separation of 150 Hz to 500 Hz from other stations. This range of separation ensures that the narrow bandwidth of CW signals doesn't overlap with adjacent frequencies, reducing the risk of interference. The closer you are to another station's frequency, the more likely your signal will interfere with their communications.

When selecting a CW transmitting frequency, what minimum separation from other stations should be used to minimize interference to stations on adjacent frequencies? (G2B04)

A. 5 Hz to 50 Hz
B. 150 Hz to 500 Hz
C. 1 kHz to 3 kHz
D. 3 kHz to 6 kHz

This principle helps to keep CW transmissions clear and allows multiple stations to operate efficiently within a shared band. Operators are encouraged to monitor the band carefully, ensuring they are not too close to other signals before transmitting. Keeping this spacing in mind ensures smoother operation on the CW bands, helping you stay within the best practices of amateur radio.

Transmit Speed

The best speed to use when answering a CQ is the fastest speed you feel comfortable copying, but it should not exceed the speed of the station sending the CQ. This ensures smooth communication and prevents operators from struggling to understand or follow the message. When in doubt, match the speed of the other operator.

What is the best speed to use when answering a CQ in Morse code? (G2C05)

A. The fastest speed at which you are comfortable copying, but no slower than the CQ
B. The fastest speed at which you are comfortable copying, but no faster than the CQ
C. At the standard calling speed of 10 wpm
D. At the standard calling speed of 5 wpm

The best speed to use when answering a CQ in Morse code is to match the sending speed of the calling station. This ensures clear communication, avoids confusion, and shows courtesy by operating at a pace the other operator is comfortable with. Adapting your speed is key to effective CW operation.

Communication Protocols

When CW operators send "KN" at the end of a transmission, they listen only for a specific station or stations. This helps control communication flow by clarifying that the operator is not open to general calls.

These CW operational standards maintain clarity, precision, and etiquette, fostering smooth and respectful communication between operators.

What does it mean when a CW operator sends "KN" at the end of a transmission? (G2C03)

A. No US stations should call
B. Operating full break-in
C. Listening only for a specific station or stations
D. Closing station now

Prosigns

A prosign (short for procedural sign) is a combination of Morse code letters that represent a specific operational command or instruction during CW (Continuous Wave) or Morse code transmissions. Unlike regular characters or abbreviations, prosigns do not have distinct spacing between the letters; they are run together as a single entity. Prosigns are used to streamline communication and improve efficiency by providing standard, quick signals that convey particular actions or procedures.

What prosign is sent to indicate the end of a formal message when using CW? (G2C08)

A. SK
B. BK
C. AR
D. KN

The prosign "AR" signals the end of a formal message in CW. This lets the receiving station know that the transmission is complete and that there is no further message content.

Prosigns streamline communication by providing universally recognized signals that operators use to maintain clarity and professionalism during CW exchanges.

Proosign	Meaning
AR	End of message
SK	End of contact; signing off
KN	Go ahead, specific station only
BK	Break; inviting another station to respond
BT	Separator between parts of a message
CQ	Calling all stations; general call
AS	Wait; stand by
RR	All received; understood
K	Go ahead; inviting any station to transmit
SOS	Distress signal; emergency communication

These prosigns are sent as a single, continuous character (no spaces between dots and dashes) and are internationally recognized, making them important for efficient Morse code operation. Operators use these to convey instructions clearly and quickly during CW exchanges.

Signal Quality and Reporting

When sending CW, what does a "C" mean when added to the RST report? (G2C07)

A. Chirpy or unstable signal
B. Report was read from an S meter rather than estimated
C. 100 percent copy
D. Key clicks

When a "C" is added to the RST report, it indicates a chirpy or unstable signal. This helps inform the transmitting station of a frequency instability or issue with their equipment, enabling them to make necessary adjustments for a cleaner signal. This is often due to power supply instability or transmitter issues. While a chirp doesn't necessarily prevent communication, it can make the signal harder to copy and is considered a flaw. Adding "C" informs the sender about this issue, allowing them to check and adjust their equipment if needed.

Advanced Techniques

Which of the following describes full break-in CW operation (QSK)? (G2C01)

A. Breaking stations send the Morse code prosign "BK"
B. Automatic keyers, instead of hand keys, are used to send Morse code
C. An operator must activate a manual send/receive switch before and after every transmission
D. Transmitting stations can receive between code characters and elements

In full break-in (QSK) CW operation, transmitting stations can receive between code characters and elements, allowing them to hear the other station while transmitting. This makes for more efficient and conversational CW exchanges because operators can interrupt or respond more quickly, improving communication flow.

Reverse Sideband

What is the benefit of using the opposite or "reverse" sideband when receiving CW? (G4A02)

A. Interference from impulse noise will be eliminated
B. More stations can be accommodated within a given signal passband
C. It may be possible to reduce or eliminate interference from other signals
D. Accidental out-of-band operation can be prevented

Using the opposite or "reverse" sideband when receiving a continuous wave (CW) signal can help reduce interference from nearby signals. CW transmissions are narrowband; when receiving them, the radio's bandwidth plays a critical role. Interference might come from signals in the adjacent sideband (either upper or lower), which can make it harder to copy the intended signal.

By switching to the opposite sideband, you are effectively tuning in a slightly different part of the spectrum, often moving away from interfering signals that might be present on the current sideband. This allows for clearer reception of the desired signal, as it reduces the likelihood of overlapping interference from nearby stations. This simple technique can significantly improve signal clarity in crowded band conditions.

Equipment

Electronic Keyer

What is the function of an electronic keyer? (G4A10)

A. Automatic transmit/receive switching
B. Automatic generation of dots and dashes for CW operation
C. To allow time for switching the antenna from the receiver to the transmitter
D. Computer interface for PSK and RTTY operation

An electronic keyer is a device used in CW operation that automatically generates the correct sequence of dots (short signals) and dashes (long signals). It simplifies the process for operators by making sending code at consistent speeds easier. Rather than manually creating each dot or dash, the operator must press a paddle, and the keyer handles the timing.

This helps reduce operator fatigue and improves accuracy, especially at higher speeds. With the electronic keyer, sending CW becomes smoother and more efficient, as it helps maintain proper spacing between characters and reduces errors in manual timing. It's a useful tool for many CW operators, especially those participating in contests or long transmissions.

Chapter Summary

CW operation is a timeless and invaluable skill in amateur radio, offering reliability and effectiveness in various conditions. You can communicate more efficiently and avoid interference by mastering techniques like zero beating, choosing the right transmitting frequency, and adjusting your sending speed. Understanding advanced practices such as full break-in (QSK) operation and using electronic keyers further enhances your CW capabilities, making communication smoother and more enjoyable.

As you continue to develop your CW skills, remember that practice and patience are key to becoming a proficient operator. By applying the concepts and techniques covered in this chapter, you'll be well-prepared to make the most of this versatile mode and enjoy the unique opportunities it offers.

Chapter 23

Digital

IMAGINE SITTING AT YOUR ham radio station, but instead of speaking into a microphone, you are typing messages on a keyboard. Your computer converts these messages into audio tones that are transmitted over the airwaves. This is the essence of digital communication, a modern advancement in ham radio that has revolutionized how operators interact. Unlike traditional voice communication, digital modes offer improved signal clarity, enhanced data transmission capabilities, and efficient bandwidth use, making them indispensable in today's amateur radio operations.

Introduction to Digital Communications

Digital communication has revolutionized amateur radio by enabling efficient, reliable, and innovative ways to exchange information. Unlike analog modes, digital modes use software and precise modulation techniques to encode data, allowing hams to operate effectively in a variety of conditions, including low power and weak signals. Understanding digital communication starts with grouping modes by their underlying modulation techniques and practical applications, giving operators a solid foundation for making informed choices.

Group 1: Narrowband Efficiency: PSK31, or Phase Shift Keying at 31.25 baud, is a digital mode known for its efficiency and minimal bandwidth usage. Its narrow bandwidth of about 31 Hz allows multiple signals to coexist without interference, making it perfect for low-power (QRP) operations and casual QSOs. PSK31's simplicity and efficiency make it a favorite among operators focused on clear communication in crowded band conditions, particularly during digital contests or for general chit-chat with other hams.

Group 2: Text-Based Reliability: Radioteletype (RTTY) is one of the oldest digital modes that is still in active use. Utilizing Frequency Shift Keying (FSK), it transmits text messages with distinct frequencies representing binary data. RTTY's robustness makes it invaluable for long-distance and emergency communications, where reliability is paramount. Operating at a standard baud rate (often 45.45 baud), it handles challenging conditions with ease, bridging decades of tradition with modern functionality.

Group 3: Weak Signal Prowess: FT8 and JT65 are standout digital modes designed for weak-signal environments.

- **FT8** shines with its 15-second transmission cycles, enabling quick exchanges under poor propagation conditions. It's particularly popular for DXing, where operators aim to establish contacts over vast distances using minimal power. FT8's highly automated and efficient nature makes it the go-to choice for operators pushing the limits of their station capabilities.

- **JT65** operates with longer, 60-second transmission cycles and incorporates advanced error-correction algorithms to decode signals often inaudible to the human ear. This mode's slower pace makes it ideal for deliberate contacts and operators with lower power setups, maintaining reliable communication even in noisy environments.

Group 4: Versatile Voice Communication: Digital voice modes such as DMR (Digital Mobile Radio), D-STAR (Digital Smart Technologies for Amateur Radio), and System Fusion bring modern communication capabilities to amateur radio. These modes combine the clarity of digital audio with advanced networking features, enabling operators to link local repeaters to global networks seamlessly.

- **DMR**: Widely adopted in both amateur and commercial radio, DMR uses Time-Division Multiple Access (TDMA) technology to split a single frequency into two separate time slots, doubling its efficiency. DMR networks, such as BrandMeister, connect operators worldwide, making it a top choice for hams looking to explore international contacts.

- **D-STAR**: Developed specifically for amateur radio, D-STAR offers integrated voice and data capabilities, along with features like callsign-based routing. Operators can transmit text, GPS coordinates, or even files alongside their voice communication, making it a versatile mode for emergency communication and experimentation.

- **System Fusion**: Known for its ease of use, System Fusion allows operators to switch seamlessly between digital and analog communication. Yaesu's proprietary C4FM technology ensures high-quality audio and straightforward integration with analog repeaters, making it an excellent option for transitioning to digital modes without leaving analog users behind.

However, digital communication also presents certain challenges. One of the main hurdles is the need for additional equipment and software. Unlike traditional voice communication, which only requires a transceiver and microphone, digital modes necessitate a computer interface and specialized software. This setup can be complex for beginners, requiring a learning curve to understand and configure the various components. Additionally, troubleshooting digital communication issues can be more challenging, as problems can arise from both hardware and software configurations.

Despite these challenges, the benefits of digital communication make it a valuable addition to any ham radio operator's skill set. By embracing digital modes, you can enhance your communication capabilities, achieve reliable contacts under poor conditions, and explore new and exciting aspects of amateur radio. As you upgrade to your General Class license, understanding digital communication will open up a world of possibilities, allowing you to connect with operators around the globe using the latest advancements in radio technology.

Digital Modes and Frequencies

Digital modes are at the forefront of modern amateur radio, offering unique advantages in clarity, efficiency, and versatility. Whether you're experimenting with FT8's weak-signal prowess or exploring the bandwidth-efficient world of RTTY, understanding these modes is critical for today's ham operators. In this section, we'll examine the key digital voice modes and where you can commonly find them on the bands, setting the stage for deeper exploration of their applications and techniques.

DMR, D-STAR, and System Fusion are digital voice modes that enable high-quality voice transmission over radio frequencies. These systems convert voice into digital data before transmitting, allowing for clearer signals and added features like GPS tracking or text messaging. Each mode operates with slightly different technology and protocols, but all offer the benefits of digital transmission. DMR is popular for its widespread use in commercial systems, while D-STAR and System Fusion are designed specifically for amateur radio.

Which of the following provide digital voice modes? (G8C16)

A. WSPR, MFSK16, and EasyPAL
B. FT8, FT4, and FST4
C. Winlink, PACTOR II, and PACTOR III
D. DMR, D-STAR, and SystemFusion

Frequencies

In what segment of the 20-meter band are most digital mode operations commonly found? (G2E08)

A. At the bottom of the slow-scan TV segment, near 14.230 MHz
B. At the top of the SSB phone segment, near 14.325 MHz
C. In the middle of the CW segment, near 14.100 MHz
D. Between 14.070 MHz and 14.100 MHz

On the 20-meter band, most digital mode operations are concentrated between 14.070 MHz and 14.100 MHz. This band portion is widely used for various digital modes, including PSK31, FT8, and other data transmissions. These modes are popular because they efficiently handle weak signals, making them ideal for long-distance communications under challenging conditions. Digital modes are less affected by noise and poor propagation, allowing operators to connect with stations worldwide even when conditions aren't ideal for voice transmission. Understanding the common digital mode frequencies can help operators quickly find contacts and take full advantage of these powerful communication tools.

Additionally, FT8 has specific frequency ranges where operators commonly use it, such as between 14.074 MHz and 14.077 MHz in the 20-meter band. Awareness of these frequencies helps you quickly tune in to FT8 activity and avoid interfering with other signals.

Which of the following is a common location for FT8? (G2E15)

A. Anywhere in the voice portion of the band
B. Anywhere in the CW portion of the band
C. Approximately 14.074 MHz to 14.077 MHz
D. Approximately 14.110 MHz to 14.113 MHz

FT8 is a low-power, highly efficient digital mode. Accurate computer timing is crucial for successful operation, and following good practices when selecting a transmission frequency helps ensure clear communication.

RTTY (Radio Teletype) is a digital mode used to transmit text-based messages over radio. One key element of RTTY is the frequency shift, which is the difference in frequency between the two tones used to represent digital data (known as mark and space). In amateur HF bands, the most common frequency shift is 170 Hz. This shift is optimal for standard RTTY operation because it balances signal robustness and bandwidth efficiency, allowing for clear communication without excessive spectrum.

What is the most common frequency shift for RTTY emissions in the amateur HF bands? (G2E06)

A. 85 Hz
B. 170 Hz
C. 425 Hz
D. 850 Hz

Understanding frequency shift is key for successful RTTY operation. In this mode, two frequencies are used to differentiate between binary signals. The most common shift used in amateur HF bands is 170 Hz, which helps ensure that RTTY signals remain narrow enough to avoid interference while being easily decoded by receiving stations.

Practical Operation and Troubleshooting

Once you've chosen a digital mode to operate, the next step is mastering its practical use. Whether it's fine-tuning your setup for FT8, understanding the nuances of RTTY, or troubleshooting a tricky signal, good operating practices are the foundation of successful communication. This section will guide you through the essentials, from selecting the right frequencies to diagnosing common decoding issues, ensuring you're ready to tackle any digital mode challenge.

FT8

FT8 is a popular digital mode known for its efficiency in low signal-to-noise ratio environments. To ensure accurate communication when using FT8, the operator's computer time must be synchronized within approximately one second. FT8 exchanges are time-sensitive, as each transmission is broken into alternating 15-second time slots. If your computer's clock is inaccurate, it can cause transmission errors or missed responses. This synchronization can be easily managed by using time-sync software to keep the computer's clock accurate.

Which of the following is required when using FT8? (G2E07)

A. A special hardware modem
B. Computer time accurate to within approximately 1 second
C. Receiver attenuator set to -12 dB
D. A vertically polarized antenna

When answering a CQ call on FT8, finding a clear frequency and transmitting during the alternate time slot from the calling station is considered good practice. This ensures that transmissions don't overlap, which can cause interference.

Which of the following is good practice when choosing a transmitting frequency to answer a station calling CQ using FT8? (G2E04)

A. Always call on the station's frequency
B. Call on any frequency in the waterfall except the station's frequency
C. Find a clear frequency during the same time slot as the calling station
D. Find a clear frequency during the alternate time slot to the calling station

---

Radio Teletype (RTTY)

What could be wrong if you cannot decode an RTTY or other FSK signal even though it is apparently tuned in properly? (G2E14)

A. The mark and space frequencies may be reversed
B. You may have selected the wrong baud rate
C. You may be listening on the wrong sideband
D. All these choices are correct

In RTTY and other Frequency Shift Keying (FSK) modes, decoding errors can happen even if the signal seems to be correctly tuned. Several factors could be at play. First, the mark and space frequencies—the two distinct tones representing binary 0s and 1s—might be reversed, which prevents accurate decoding. Another issue could be that you've selected the wrong baud rate, meaning the data transmission speed doesn't match between the sender and receiver. Listening on the wrong sideband(LSB vs. USB) could also distort the signal, as these modes require the correct sideband for proper reception.

Sideband Usage

Digital modes utilize different methods for transmitting data, such as AFSK (Audio Frequency Shift Keying), and they rely on specific sidebands depending on the type of transmission. Understanding the proper sideband to use ensures that your signals are transmitted correctly and can be easily received and decoded by other operators. In amateur radio, RTTY (Radio Teletype) signals and digital modes like FT8, JT65, and others have their own standard operating conventions regarding sideband usage.

Which mode is normally used when sending RTTY signals via AFSK with an SSB transmitter? (G2E01)

A. USB
B. DSB
C. CW
D. LSB

RTTY via AFSK with an SSB transmitter typically uses LSB (Lower Sideband). This is a common practice for sending RTTY signals, especially in the HF bands, where the lower sideband is the norm below 10 MHz. AFSK is a method where audio tones are modulated onto a carrier by using an SSB transmitter.

What is the standard sideband for JT65, JT9, FT4, or FT8 digital signal when using AFSK? (G2E05)

A. LSB
B. USB
C. DSB
D. SSB

For modes like JT65, JT9, FT4, or FT8, the standard sideband used when transmitting via AFSK is USB (Upper Sideband). This distinction is important because most modern digital modes on HF and higher frequencies use USB as the standard, even in frequency ranges where LSB would typically be used for voice communications. Knowing which sideband to use for each digital mode ensures compatibility with other stations and helps avoid unnecessary interference.

By understanding these sideband conventions, operators can ensure their signals are correctly transmitted and received, optimizing their chances of making successful digital mode contacts.

Advanced Digital Communication Protocols

Digital communication protocols like PACTOR, VARA, and Winlink offer hams powerful tools for reliable and efficient data transmission. These protocols excel in delivering text, email, and other data in both routine and emergency scenarios, even under challenging conditions. Here, we'll explore how these systems work, how to use them effectively, and their critical roles in modern amateur radio operations.

PACTOR

PACTOR is a digital communication mode designed for robust and error-free data exchange over HF radio. One of the key features of the PACTOR protocol is that it establishes a connection between only two stations at a time, meaning that joining an existing conversation or connection is not possible. This limitation ensures that the data transfer remains efficient and secure, with minimal interference or data loss. If you come across an active PACTOR session between two stations, the protocol does not allow for a third station to participate or interrupt the exchange.

How do you join a contact between two stations using the PACTOR protocol? (G2E09)

A. Send broadcast packets containing your call sign while in MONITOR mode
B. Transmit a steady carrier until the PACTOR protocol times out and disconnects
C. Joining an existing contact is not possible, PACTOR connections are limited to two stations
D. Send a NAK code

PACTOR is designed for reliable two-way communication between two stations only. This ensures secure, private, and interference-free data transfer, which makes it popular for applications like email transmission over HF or maritime communications. Because of this limitation, attempting to join an ongoing PACTOR connection is impossible, reinforcing this protocol's efficiency and focus.

Winlink

Winlink is a global messaging system that allows users to send and receive emails over HF, VHF, or UHF radio frequencies. It is particularly useful for emergency communications and off-grid scenarios where the internet is unavailable. Winlink can bridge traditional radio communication with the internet, allowing users to exchange messages with regular email addresses, even from remote locations.

Here's a detailed breakdown of how Winlink works and its key components:

How Winlink Works:

- **Radio-Based Email System**: At its core, Winlink allows users to send and receive emails over amateur radio frequencies using specialized software on the internet. These emails can be sent peer-to-peer via radio or routed to the internet via Radio Message Servers (RMS) stations.

- **Message Routing**: Messages (packets) sent over Winlink can either be relayed directly to another radio operator or sent to an RMS station connected to the Internet. The RMS server forwards the email to the recipient via the Internet.

- **Multiple Bands**: Winlink operates as a wireless network over HF, VHF, and UHF frequencies. On HF, long-distance communication is possible via ionospheric propagation, allowing users to connect to RMS stations that may be far away geographically.

- **Hybrid Network**: Winlink's hybrid network is unique. Even if the internet is down, Winlink can continue to function by relaying messages via radio from one RMS station to another. This makes it especially useful for emergency communication in disaster-prone areas.

- **How to Establish Contact:** To establish contact with a digital messaging system gateway station, such as those used for Winlink, you must transmit a "connect" message on the station's published frequency. Digital gateways are often used to send and receive messages in modes like PACTOR, VARA, or ARDOP. These gateways act as intermediaries, facilitating digital data transfer (such as emails) between radio operators and internet-linked servers. The "connect" message lets the gateway know that you are attempting to establish a communication link. The process is similar to dialing into an internet server in early internet systems, where the right frequency and digital protocol are crucial for a successful connection.

Based on that info, let's answer some exam questions.

Which of the following describes Winlink? (G2E12)

A. An amateur radio wireless network to send and receive email on the internet
B. A form of Packet Radio
C. A wireless network capable of both VHF and HF band operation
D. All of the above

If 'A,' 'B,' and 'C' are true, then it must be 'all of the above.'

Which of the following is a way to establish contact with a digital messaging system gateway station? (G2E10)

A. Send an email to the system control operator
B. Send QRL in Morse code
C. Respond when the station broadcasts its SSID
D. Transmit a connect message on the station's published frequency

---

Software and Hardware:

- **Winlink Software**: To use the system, users need software like Winlink Express, which can connect to the Winlink system through their computer. This software handles composing, sending, and receiving emails, similar to traditional email programs.

- **Modems**: Winlink uses modems (or sound cards) that convert digital email data into radio signals that can be transmitted over the airwaves. Examples include PACTOR and VARA modems for HF communication.

- **Winlink Remote Message Server**: These are the stations that provide the link between amateur radio frequencies and the internet. They act as relay points, sending messages from the radio network to the internet-based email systems and vice versa. They are also known as Gateways.

Question time!

What is another name for a Winlink Remote Message Server? (G2E13)

A. Terminal Node Controller
B. Gateway
C. RJ-45
D. Printer/Server

Winlink Modes:

- **Peer-to-Peer Mode**: In this mode, emails are sent directly from one amateur radio station to another without using the internet or RMS gateways.

- **RMS Mode**: This is the most common method. The message is sent via an RMS gateway, which forwards it to the recipient's internet-based email address.

- **Hybrid Mode**: In areas where internet connectivity is disrupted, the hybrid mode allows the system to relay messages entirely over the radio from one RMS to another until it reaches a station with internet access.

Winlink Applications:

- **Emergency Communications (EMCOMM)**: Winlink is an invaluable tool in disaster response. It can exchange messages, share forms (like ICS forms for incident management), and coordinate relief efforts when normal communication infrastructure is unavailable.

- **Off-Grid Communications**: Winlink allows users in remote areas without internet access to stay connected with the world via email.

- **Sailing and Remote Expeditions**: Winlink is popular among sailors and adventurers in remote locations because it allows them to send and receive emails even when they are far from any cellular network.

Winlink Protocols:

- **PACTOR** is one of the most efficient protocols used in Winlink. It is known for high-speed and reliable data transfer over HF radio. PACTOR modems can operate under noisy and weak signal conditions.

- **VARA**: <u>VARA is a sound card digital protocol used with Winlink</u> that offers fast transmission speeds and is an alternative to PACTOR, though PACTOR tends to offer higher reliability.

- **ARDOP**: This modern protocol designed specifically for Winlink offers a flexible and open-source solution for digital communication over HF.

- **AX.25 Packet Radio**: Used for VHF/UHF Winlink operation, this protocol is slower but efficient for shorter-range communications.

What is VARA? (G2E02)

A. A low signal-to-noise digital mode used for EME (moonbounce)
B. A digital protocol used with Winlink
C. A radio direction finding system used on VHF and UHF
D. A DX spotting system using a network of software defined radios

---

Benefits and Challenges:

- **Benefits**: Winlink is highly adaptable and resilient in disaster situations, providing a vital link for communication when other systems fail. Its ability to connect via HF radio makes it an excellent tool for long-distance communication, even without the internet.

- **Challenges**: Setting up Winlink requires some technical expertise, including understanding how to interface radios with modems and configuring the necessary software. Licensing and compliance with amateur radio rules are also essential for proper operation.

- **Interference**: When using digital modes like PACTOR or VARA, interference from other signals can cause various issues with your transmission. The most common symptoms include <u>frequent retries or timeouts</u>, which occur when the signal is disrupted to the point that the system must repeatedly try to send the data. You might also notice <u>long pauses in message transmission</u> as the system waits for a clear path to continue the transmission. Finally, severe interference may lead to the <u>failure to establish a connection between stations</u> entirely, preventing any communication. This happens because PACTOR and VARA are sensitive to signal clarity, and interference from overlapping frequencies or noise can significantly impact their effectiveness.

What symptoms may result from other signals interfering with a PACTOR or VARA transmission? (G2E03)

A. Frequent retries or timeouts
B. Long pauses in message transmission
C. Failure to establish a connection between stations
D. All these choices are correct

Winlink bridges radio communications and the internet, making it a versatile tool for emergency communication, off-grid messaging, and international coordination. It leverages both traditional radio frequencies and the internet to ensure messages get through under nearly any circumstances, making it a critical system for ham radio operators involved in emergency preparedness, remote expeditions, and disaster recovery.

Digital Networking and Emergency Applications

The integration of digital networking into amateur radio has revolutionized emergency communication and expanded the scope of what hams can achieve. Technologies like the Amateur Radio Emergency Data Network (AREDN) provide mesh networking solutions, while shared spectrum with unlicensed Wi-Fi enables innovative uses of existing equipment. This section will explore these exciting applications, highlighting their relevance to emergency preparedness and technical exploration.

Amateur Radio Emergency Data Network

An Amateur Radio Emergency Data Network (AREDN) mesh network is a high-speed data network designed to be deployed during emergencies or community events when regular communication systems are overloaded or unavailable. These networks use amateur radio frequencies and equipment to set up local area networks that can provide a variety of services, such as email, VoIP (Voice over IP) communications, file sharing, and even live video streaming. The network's decentralized and flexible nature allows for quick deployment, with each node in the network able to route data efficiently, even if some nodes go down. This makes it an invaluable tool for emergency responders and amateur radio operators involved in disaster relief.

What is the primary purpose of an Amateur Radio Emergency Data Network (AREDN) mesh network? (G2E11)

A. To provide FM repeater coverage in remote areas
B. To provide real-time propagation data by monitoring amateur radio transmissions worldwide
C. To provide high-speed data services during an emergency or community event
D. To provide DX spotting reports to aid contesters and DXers

---

The primary purpose of an AREDN mesh network is to deliver high-speed data services during emergencies when traditional communication networks fail. By using amateur radio frequencies, these networks can provide reliable and scalable communication options, ensuring critical data gets through.

Spectrum Use

Amateur radio operators share certain parts of the radio frequency spectrum with other services, including unlicensed users like Wi-Fi networks. The 2.4 GHz band is a shared range where licensed amateur radio and unlicensed services, such as Wi-Fi, Bluetooth, and other short-range devices, operate. Understanding which frequency bands are shared helps radio operators avoid interference with other services and ensures smooth operation for everyone using the spectrum.

On what band do amateurs share channels with the unlicensed Wi-Fi service? (G8C01)

A. 432 MHz
B. 902 MHz
C. 2.4 GHz
D. 10.7 GHz

Amateur radio operators share the 2.4 GHz band with unlicensed services, including Wi-Fi. The 2.4 GHz band is widely used for Wi-Fi networks because it offers a good balance between range and data-carrying capacity. While hams can legally transmit on this frequency, they must do so with caution to avoid causing interference to unlicensed users, as Wi-Fi and other devices also rely on this band for communication.

As an amateur radio operator, knowing about this shared usage is important to ensure harmonious operation with other users, such as Wi-Fi.

Software-Defined Radio (SDR)

Software-Defined Radio (SDR) represents a groundbreaking evolution in ham radio technology. Unlike traditional radios that rely on fixed hardware components, SDRs perform core functions like mixing, filtering, and modulating through software, offering unparalleled flexibility and adaptability. With the ability to update and reconfigure systems via software, SDRs are future-proof and versatile.

One of SDRs' most valuable features is real-time spectrum analysis and waterfall displays, which allow operators to visualize and analyze signals across multiple bands, from HF to UHF. These tools are indispensable for detecting interference, optimizing operating conditions, and experimenting with digital modes like FT8 and PSK31. SDRs also enhance signal processing through digital filters, noise reduction, and advanced automation, making them powerful tools for both beginners and experienced operators.

Setting up an SDR involves selecting compatible hardware, such as the SDRplay RSP1A or HackRF One, and pairing it with software like SDR# or HDSDR. Once configured, these systems open the door to monitoring faint signals, exploring new modes, and enhancing communication capabilities in ways that traditional radios cannot match.

By integrating cutting-edge technology with the spirit of amateur radio experimentation, SDRs empower operators to stay at the forefront of innovation, making them an essential tool for the modern ham.

Software-Defined Radios (SDR) and Modulation

Software-defined radios (SDR) represent a significant evolution in radio operation. They rely on software rather than solely on hardware components to process radio signals. This allows SDR systems to be much more flexible and adapt to different frequencies, modulation types, and other parameters through programming rather than physical changes. One key concept in SDR is I-Q modulation, where a signal's in-phase (I) and quadrature (Q) components are processed for efficient modulation and demodulation.

What is the phase difference between the I and Q RF signals that software-defined radio (SDR) equipment uses for modulation and demodulation? (G7C09)

A. Zero
B. 90 degrees
C. 180 degrees
D. 45 degrees

In software-defined radios, I-Q modulation is a method where two signals, one in-phase (I) and the other in quadrature (Q), represent different components of the radio signal. The key to this modulation technique is the 90-degree phase difference between the I and Q signals. This allows for the creation of complex modulation schemes that can be easily processed and demodulated by the SDR.

What is an advantage of using I-Q modulation with software-defined radios (SDRs)? (G7C10)

A. The need for high resolution analog-to-digital converters is eliminated
B. All types of modulation can be created with appropriate processing
C. Minimum detectible signal level is reduced
D. Automatic conversion of the signal from digital to analog

I-Q modulation offers SDRs a unique advantage in that virtually any modulation scheme can be generated with the right software processing. Whether it's AM, FM, SSB, or digital modes, SDRs can use I-Q modulation to create and manipulate signals, giving them an edge in flexibility and adaptability compared to traditional radios that require different hardware for each modulation type.

Which of these functions is performed by software in a software-defined radio (SDR)? (G7C11)

A. Filtering
B. Detection
C. Modulation
D. All these choices are correct

One of the main benefits of SDR technology is the ability to shift many functions—such as filtering, signal detection, and modulation—from hardware to software. This makes SDRs incredibly versatile and upgradeable, as software can be modified or updated to improve performance or adapt to new standards without changing the hardware.

Software-defined radios (SDRs) revolutionize radio technology by relying on software to perform functions traditionally managed by hardware. Using I-Q modulation techniques and advanced digital signal processing, SDRs provide unparalleled flexibility, allowing operators to easily switch between modulation modes and frequencies. As software improves, so do the capabilities of these radios, making them powerful tools for modern communications.

Chapter Summary

Incorporating digital communication into your operations requires careful planning and setup. By ensuring you have the right equipment, properly configuring your interfaces, and addressing any setup issues, you can enjoy the benefits of digital modes, enhance your communication capabilities, and connect with operators worldwide.

Chapter 24

Interference

INTERFERENCE IS AN INEVITABLE challenge in ham radio operations, affecting signal clarity and communication quality. From household devices to power lines, a wide range of sources can disrupt your transmissions, making it essential for operators to understand and address these issues effectively. Learning interference mitigation enhances your ability to communicate reliably and ensures compliance with regulations and respect for other operators on the airwaves.

In this chapter, we'll explore the most common types of interference, their causes, and practical solutions. You'll learn to identify specific interference patterns, use tools like bypass capacitors and ferrite chokes, and prevent issues like ground loops. Whether diagnosing distorted signals or eliminating RF noise from your setup, this chapter provides the tools and knowledge to keep your station operating at its best.

Several techniques and components can help reduce unwanted noise when dealing with RF interference in audio systems. One common solution is using a bypass capacitor. Bypass capacitors work by shunting high-frequency RF signals to the ground, preventing them from traveling along the audio path and causing interference. This is especially useful when the RF signals are coupled into sensitive circuits, such as amplifiers or microphones, where they can cause unwanted noise or distortion.

Which of the following might be useful in reducing RF interference to audio frequency circuits? (G4C01)

A. Bypass inductor
B. Bypass capacitor
C. Forward-biased diode
D. Reverse-biased diode

Another potential source of RF interference is arcing at a poor electrical connection. Arcing can generate noise across a wide range of frequencies and is often caused by loose or corroded connections in electrical systems. This type of interference can be challenging to diagnose because it covers multiple frequencies, but addressing the faulty connection can eliminate the problem.

Which of the following could be a cause of interference covering a wide range of frequencies? (G4C02)

A. Not using a balun or line isolator to feed balanced antennas
B. Lack of rectification of the transmitter's signal in power conductors
C. Arcing at a poor electrical connection
D. Using a balun to feed an unbalanced antenna

One effective solution for interference caused by common-mode currents on audio cables is to place a ferrite choke on the cable. A ferrite choke helps block high-frequency signals traveling along the outside of the cable, reducing interference without affecting the audio signal. This is a common solution when RF signals are being picked up by long audio cables acting as unintended antennas.

Which of the following would reduce RF interference caused by common-mode current on an audio cable? (G4C08)

A. Place a ferrite choke on the cable
B. Connect the center conductor to the shield of all cables to short circuit the RFI signal
C. Ground the center conductor of the audio cable causing the interference
D. Add an additional insulating jacket to the cable

To combat RF interference, use a bypass capacitor to filter out high-frequency noise, correct poor electrical connections that may be causing arcing, and add ferrite chokes to audio cables to eliminate common-mode currents. These steps help to maintain clear audio quality and prevent interference from external RF signals.

By combining these methods, amateur radio operators and audio engineers can reduce or eliminate the impact of RF interference on their audio equipment.

Interference Sounds

When operating a ham radio station, it's not uncommon to encounter different types of sounds and interference that can affect your transmission and reception quality. These sounds can be clues to specific types of interference, helping you diagnose and fix potential issues in your setup.

Each type of interference produces a unique audio signature, from distorted voices to humming or clicking sounds, making it easier to identify and address the underlying problems. This section will explore the different types of interference and associated characteristic sounds, providing a guide to recognizing and troubleshooting these common issues in ham radio setups.

What sound is heard from an audio device experiencing RF interference from a single sideband phone transmitter? (G4C03)

A. A steady hum whenever the transmitter is on the air
B. On-and-off humming or clicking
C. Distorted speech
D. Clearly audible speech

When RF interference from a single sideband (SSB) phone transmitter affects an audio device, it typically results in distorted speech. This occurs because the SSB signal, which carries voice information, is unintentionally picked up by the audio equipment, causing it to become distorted as the RF signal interferes with normal audio processing. This type of interference can make voices sound garbled or unclear, affecting communication quality in radio setups or audio systems.

CW Transmitter

> **What sound is heard from an audio device experiencing RF interference from a CW transmitter? (G4C04)**
>
> **A. On-and-off humming or clicking**
> B. A CW signal at a nearly pure audio frequency
> C. A chirpy CW signal
> D. Severely distorted audio

In the case of CW (Continuous Wave) interference, the sound heard in an audio device is usually an on-and-off humming or clicking. This happens because CW transmissions are simple, on-and-off keyed RF signals, so when an audio device inadvertently picks them up, they manifest as repetitive pulses, clicks, or hums that follow the CW transmission pattern.

Ground Loop

> **What could be a symptom caused by a ground loop in your station's audio connections? (G4C10)**
>
> **A. You receive reports of "hum" on your station's transmitted signal**
> B. The SWR reading for one or more antennas is suddenly very high
> C. An item of station equipment starts to draw excessive amounts of current
> D. You receive reports of harmonic interference from your station

A ground loop in an audio setup, which occurs when there are multiple ground paths between devices, can result in a persistent "hum" that is reported by other stations during transmission. This hum is caused by electrical interference being picked up and fed into the audio system, making the transmitted signal less clear.

Summary of Audio Device RF Interference:

- **SSB transmitter interference**: Results in distorted speech in audio devices.

- **CW transmitter interference**: Produces on-and-off humming or clicking sounds.

- **Ground loops**: Can lead to a persistent hum in transmitted signals.

By understanding these symptoms, operators can better diagnose and correct RF interference issues in their audio and radio setups.

Chapter Summary

Interference is an inevitable part of ham radio operations, but understanding its causes and solutions is key to maintaining a clear and reliable signal. By identifying the sources of interference—whether it's RF noise from household devices, arcing electrical connections, or common-mode currents—and applying targeted fixes such as bypass capacitors, ferrite chokes, or correcting ground loops, operators can significantly reduce its impact.

Recognizing the unique sounds and symptoms of interference further aids in swift diagnosis and resolution. Mastering these techniques not only enhances your operating experience but also ensures compliance with communication standards, allowing for more enjoyable and effective transmissions.

Rules & Regulations

IMAGINE A VAST, OPEN field where you can only explore a small corner. Almost the entire field is open to you with your General Class license. This newfound freedom brings exciting opportunities and responsibilities. As a General Class operator, you gain access to additional bands and frequencies that were previously out of reach. This chapter will guide you through your new frequency privileges and regulations, helping you make the most of them and understand the nuances that come with your upgraded license.

Chapter 25

Frequency Privileges and Band Plans

ITU Regions

THE INTERNATIONAL TELECOMMUNICATION UNION (ITU) divides the world into three regions to manage and regulate global radio frequency usage. These regions help coordinate radio communication practices and frequency allocations to ensure efficient and non-interfering use of the radio spectrum. Different regions have distinct frequency allocations, and understanding the applicable region is crucial for amateur operators to follow the correct regulations.

For radio amateurs operating in North and South America, the frequency allocations of ITU Region 2 apply. This region includes the Western Hemisphere, and its frequency allocations differ from those of Region 1 (Europe, Africa, and the Middle East) and Region 3 (Asia and the Pacific). Region 2 allocations help ensure consistent communication practices across North and South America while coordinating with the ITU to avoid user interference in other regions.

The frequency allocations of which ITU region apply to radio amateurs operating in North and South America? (G1E06)

A. Region 4
B. Region 3
C. Region 2
D. Region 1

Frequency Privileges

With a General license, you access an array of high-frequency (HF) bands that significantly expand your operating capabilities compared to the Technician Class. These bands include 160, 80, 40, 20, 15, and 10 meters, each offering unique propagation characteristics and opportunities for long-distance communication.

Beyond these HF bands, General operators also enjoy expanded privileges on VHF and UHF bands. While Technician Class licensees have access to these bands, the General Class license offers additional segments and power limits, enhancing your ability to communicate over longer distances and through various propagation modes. This broader access increases your reach and opens up new avenues for experimenting with different modes and frequencies.

The differences between Technician and General Class privileges are noteworthy. As a General Class operator, your power limits increase significantly on certain bands, allowing you to transmit with up to 1500 watts PEP on most HF bands. This is a substantial upgrade from the 200-watt limit on the Technician bands. Moreover, the General Class license grants you access to expanded band segments for voice (SSB), CW, and digital modes. This expanded access is particularly advantageous for international communication, enabling you to participate in DX contacts and join international nets and contests.

Bands

60-Meter

The 60-meter band is a unique allocation in the amateur radio spectrum, designed to operate on five specific channels (or frequencies) rather than a continuous frequency range like other HF bands. Because it is shared with federal and other government users, the FCC has implemented additional rules and restrictions to minimize interference and ensure coexistence between amateur radio and other services.

For General class licensees, the 60-meter band offers a rare opportunity to operate within a fixed channelized structure, which differs from other bands' open, frequency-sweeping operations. Here are the key points to know:

- **Channelized Operation**: The five channels are 5330.5, 5346.5, 5357.0, 5371.5, and 5403.5 kHz, and operators must ensure that their signals remain centered on these frequencies.

- **Power Limits**: Operators are limited to 100 watts PEP Effective Radiated Power (ERP). This ensures that amateur transmissions remain within acceptable limits to prevent interference with primary users on the band.

- **Antenna Regulations**: If you use an antenna other than a dipole, you are required to keep a record of its gain. The ERP calculation is based on a dipole's performance, and using a more directional or high-gain antenna can exceed the power limit without proper documentation.

- **Operating Modes**: USB (Upper Sideband) is the primary mode for voice transmissions, and other digital and CW modes are allowed as long as they fit within the channel bandwidth restrictions.

Overall, the 60-meter band provides valuable opportunities for communications, particularly in emergency and disaster response situations where operators may need to coordinate with other services. The band's unique constraints make it an interesting and important part of the General class operating privileges.

Channelized Bands

A channel in radio communication refers to a specific, pre-assigned frequency or set of frequencies within a broader radio band that is designated for communication. Instead of allowing operators to use any frequency within a range, channels are fixed, clearly defined points on the spectrum where communication is permitted. Each channel has a specific center frequency and bandwidth, ensuring that multiple operators can share the same band without interference.

In these cases, operators must tune their radios to the exact frequency of the channel and follow the band's specific operating rules.

Key Points:

- **Fixed frequencies**: A channel is a specific frequency within a band, not a range.

- **Shared use**: Channels help organize communication, especially in shared bands, by preventing users from interfering with each other.

- **Efficient spectrum use:** predefined channels ensure that users use the available spectrum efficiently and fairly.

The 60-meter band is unique in that amateur operators are limited to transmitting on specific channels rather than a continuous frequency range. Five channels are designated for amateur use in the United States, and these channels must be used exactly as specified, with no deviation. This channelization is important because 60 meters is a shared band, meaning both amateurs and government services use it. Channelization helps avoid interference and ensures that both users can coexist without disrupting communications.

Which amateur bands are restricted to communication only on specific channels rather than frequency ranges? (G1A04)

A. 11 meters
B. 12 meters
C. 30 meters
D. 60 meters

Special Regulations on the Exam

The 60-meter band is unique in amateur radio as it operates on specific channels and is shared with federal and other users. The FCC imposes additional regulations to ensure responsible use, including specific power limits and antenna guidelines. These rules help manage interference and ensure that the band is used efficiently and fairly by both amateur and non-amateur services.

When operating on the 60-meter band, if you are using an antenna other than a dipole, the FCC requires that you keep a record of the gain of your antenna. This is because power limits on the 60-meter band are measured in terms of Effective Radiated Power (ERP) relative to a dipole antenna. Keeping a record of the antenna gain ensures that you remain within the 100-watt ERP limit set for this band, even when using more complex or directional antennas that can increase your signal strength.

Which of the following is required by the FCC rules when operating in the 60-meter band? (G1C04)

A. If you are using an antenna other than a dipole, you must keep a record of the gain of your antenna
B. You must keep record of the date, time, frequency, power level, and stations worked
C. You must keep a record of all third-party traffic
D. You must keep a record of the manufacturer of your equipment, and the antenna used

This helps operators stay compliant with power restrictions and prevents interference with other users on the 60-meter band.

Bandwidth

The FCC enforces strict rules on how much bandwidth an amateur transmission can occupy to avoid interference. This ensures that while taking advantage of the 60-meter channels, amateur operators do not unintentionally interfere with other services that also use this band. The rules apply particularly to USB (Upper Sideband) voice transmissions, the primary mode used on this band.

On the 60-meter band, the maximum bandwidth permitted for amateur stations transmitting on USB frequencies is 2.8 kHz. This limitation ensures that the signals remain narrow enough to fit within the allocated channels without spilling into adjacent frequencies, which could interfere with other communications. The 2.8 kHz limit accommodates typical voice transmissions while preventing over-modulation or excessive bandwidth usage.

> **What is the maximum bandwidth permitted by FCC rules for amateur radio stations transmitting on USB frequencies in the 60-meter band? (G1C03)**
>
> **A. 2.8 kHz**
> B. 5.6 kHz
> C. 1.8 kHz
> D. 3 kHz

To remember this for the exam, think of 2.8 kHz as the standard bandwidth for voice transmissions in the USB mode on this shared band, ensuring that communication remains efficient and interference-free.

15-Meter

The 15-meter band is a valuable resource for General class licensees, offering excellent long-distance communication opportunities, especially during periods of high solar activity. As a General class operator, you gain access to a significant portion of this band, specifically from 21.275 kHz to 21.450 kHz. This range provides ample space for voice, CW (Morse code), and digital modes, allowing General licensees to make global contacts, particularly during daylight hours.

And now, for an exam question.

The frequency of 21.300 kHz falls within the General class portion of the 15-meter band. General class operators are allowed to transmit in the range from 21.275 kHz to 21.450 kHz on the 15-meter band. This portion provides general class licensees with a substantial range of communication capabilities. In contrast, Amateur Extra licensees have additional access to a smaller segment of the band. General class operators must ensure they are transmitting within this authorized range to avoid operating outside of their privileges.

> **Which of the following frequencies is within the General class portion of the 15-meter band? (G1A09)**
>
> A. 14250 kHz
> B. 18155 kHz
> **C. 21300 kHz**
> D. 24900 kHz

Remember, 21,300 kHz is part of the "middle ground" in the 15-meter band, where General-class operators have full access, while certain lower frequency segments are reserved for Amateur Extra operators.

10-Meter

The 10-meter band is unique and versatile, offering exciting opportunities for General class licensees. It sits at the upper end of the HF spectrum. It provides the potential for both local and long-distance (DX) communication, particularly during periods of high solar activity. The band is well-suited for various modes, including voice (SSB), CW, digital modes, and even repeater use. General class operators have broad access to the 10-meter band, allowing them to experiment with different modes and take advantage of favorable propagation conditions to make contacts around the world, especially when the band is open during the peak of the solar cycle.

Continuous Wave

CW (Continuous Wave) is a mode of communication used primarily for Morse code transmissions. Different license classes have specific privileges on various frequency bands, but CW operation is often permitted across broader segments because of its narrow bandwidth and efficient spectrum use. The 10-meter band is a portion of the HF spectrum that offers a wide range of privileges for General class licensees.

On the 10-meter band, stations with a General class control operator may transmit CW emissions across the entire band. Unlike some other bands where specific frequency ranges are reserved for higher license classes, General class operators have full access to the 10-meter band for CW transmissions. This makes the 10-meter band a valuable resource for General license holders who wish to use Morse code for long-distance communication.

On which amateur frequencies in the 10-meter band may stations with a General class control operator transmit CW emissions? (G1A07)

A. 28.000 MHz to 28.025 MHz only
B. 28.000 MHz to 28.300 MHz only
C. 28.025 MHz to 28.300 MHz only
D. The entire band

Repeaters

On the 10-meter band, repeaters play a crucial role in providing long-distance communication, particularly during periods of favorable propagation. Unlike VHF or UHF repeaters, which are commonly used for local communication, 10-meter band repeaters allow for extended reach, sometimes spanning hundreds or even thousands of miles.

The portion of the 10-meter band available for repeater use is above 29.5 MHz. This section of the band is specifically allocated for repeater operation, enabling General and higher-class licensees to take advantage of repeater networks for long-distance communication. Repeaters in this frequency range can provide clear, reliable communications over large distances, particularly when the band is open due to favorable propagation conditions.

What portion of the 10-meter band is available for repeater use? (G1A10)

A. The entire band
B. The portion between 28.1 MHz and 28.2 MHz
C. The portion between 28.3 MHz and 28.5 MHz
D. The portion above 29.5 MHz

Everything above this frequency is available for repeaters, helping to extend communication range beyond what is possible with direct line-of-sight radio operation.

12-Centimeters

The 12-centimeter band is part of the SHF (Super High Frequency) spectrum, spanning frequencies from 2.3 GHz to 2.45 GHz. Amateur radio operators primarily use frequencies close to 2.4 GHz. This band is shared with many other services, including Wi-Fi, Bluetooth, and other unlicensed consumer technologies that operate under Part 15 of the FCC rules.

Amateur radio operators use the 12 cm band for various communications, including short-range, line-of-sight transmissions and experiments with digital and spread spectrum techniques. One key aspect of this band is that it falls into the microwave frequency range, allowing for high data rate communications and making it ideal for fast-scan television and various digital modes.

Due to its shared nature, amateur operators must be mindful of potential interference with unlicensed devices and use proper coordination and care when operating on the 2.4 GHz portion of the band. The relatively short wavelength of about 12.5

centimeters makes it suitable for highly directional antenna designs, allowing for focused signal transmission and reception over short distances.

Amateur radio operators in the 2.4 GHz band must be mindful that this portion of the spectrum is shared with unlicensed services like Wi-Fi. While both services occupy this frequency range, amateur stations are not permitted to communicate with non-licensed Wi-Fi stations. The FCC rules keep amateur operations separate from commercial and consumer applications, ensuring each service operates within its designated scope without causing interference or blending the two services.

In what part of the 2.4 GHz band may an amateur station communicate with non-licensed Wi-Fi stations? (G1E07)

A. Anywhere in the band
B. Channels 1 through 4
C. Channels 42 through 45
D. No part

Amateur stations are not allowed to communicate with non-licensed Wi-Fi stations in any part of the 2.4 GHz band. Even though both amateur radio and Wi-Fi operate within this shared spectrum, they serve different purposes, and the rules prohibit cross-communication between them. Wi-Fi is regulated under Part 15 of the FCC rules, while amateur radio is governed by Part 97. These rules prevent service interference and maintain the integrity of amateur operations and consumer networks like Wi-Fi.

Restrictions

Amateur Extra Licensees

Higher-level licenses, like the Amateur Extra class, come with greater operating privileges. While General licensees have access to most of the HF bands, certain portions of these bands are reserved exclusively for Amateur Extra licensees. These reserved segments allow more experienced operators to enjoy less crowded operating conditions and prioritize using these critical frequencies, especially in busy bands.

The 80-meter, 40-meter, 20-meter, and 15-meter HF bands have segments that are exclusively allocated to Amateur Extra licensees. These segments are set aside to give Extra class operators exclusive access to certain frequencies within these bands, which means that General and Technician class operators cannot transmit in these areas. These reserved segments are typically at the lower end of each band, offering more experienced operators a quieter, less congested space for communication.

Which HF bands have segments exclusively allocated to Amateur Extra licensees? (G1A08)

A. All HF bands
B. 80 meters, 40 meters, 20 meters, and 15 meters
C. All HF bands except 160 meters and 10 meters
D. 60 meters, 30 meters, 17 meters, and 12 meters

To remember this for the exam, think of the 80, 40, 20, and 15-meter bands as shared spaces, but with exclusive "VIP sections" reserved for Amateur Extra operators, providing them with additional operating privileges.

Primary and Secondary Users

Certain frequency bands are shared between different services, which are classified as primary or secondary users. Primary users have priority on the band, and their communications take precedence. Secondary users, such as amateur radio operators on some bands, are allowed to operate but must ensure that they do not interfere with the primary users. Additionally, secondary users must be prepared to accept any interference caused by primary users.

When the FCC designates amateur radio operators as secondary users on a band, it means that they must not cause any harmful interference to primary users and must accept any interference from those primary users. Primary users, such as government or military services, have the right to use the frequency without disruption. Suppose a secondary user, like a ham operator, is causing interference to a primary user. In that case, the amateur operator must reduce power, change frequency, or stop transmitting.

Which of the following applies when the FCC rules designate the amateur service as a secondary user on a band? (G1A06)

A. Amateur stations must record the call sign of the primary service station before operating on a frequency assigned to that station
B. Amateur stations may use the band only during emergencies
C. Amateur stations must not cause harmful interference to primary users and must accept interference from primary users
D. Amateur stations may only operate during specific hours of the day, while primary users are permitted 24-hour use of the band

---

Partial Band Restrictions

General class licensees have significant access to the HF bands, but on some bands, they are restricted from using the entire voice (phone) segment. These restrictions help manage band usage and ensure that more experienced operators, such as Amateur Extra class licensees, have exclusive access to certain frequencies. General class operators are typically allowed to use the upper portion of the voice segment on these restricted bands, enabling them to participate in voice communication while adhering to FCC regulations.

When General class licensees are not permitted to use the entire voice portion of a band, they are allowed to operate in the upper-frequency portion of the voice segment. This means that while Amateur Extra operators may have access to the entire voice section, General class operators are limited to the higher end of that portion. These upper segments are still well-suited for voice communication and offer ample contact opportunities.

When General class licensees are not permitted to use the entire voice portion of a band, which portion of the voice segment is available to them? (G1A11)

A. The lower frequency portion
B. The upper frequency portion
C. The lower frequency portion on frequencies below 7.3 MHz and the upper portion on frequencies above 14.150 MHz
D. The upper-frequency portion on frequencies below 7.3 MHz and the lower portion on frequencies above 14.150 MHz

---

While we can't operate on the lower part of the voice range reserved for Extra class, the upper portion provides plenty of room for our communication needs.

HF and MF Band Restrictions

General licensees are granted access to a large portion of the HF and MF bands used for long-distance communication. However, some restricted segments within these bands are reserved for higher license classes.

On the 80-meter, 40-meter, 20-meter, and 15-meter bands, there are specific portions where General class licensees cannot transmit, as these segments are reserved for Amateur Extra class operators. These restrictions give more experienced operators exclusive access to certain parts of the spectrum. General class operators must be aware of these limits to avoid operating outside their privileges.

On which HF and/or MF amateur bands are there portions where General class licensees cannot transmit? (G1A01)

A. 60 meters, 30 meters, 17 meters, and 12 meters
B. 160 meters, 60 meters, 15 meters, and 12 meters
C. 80 meters, 40 meters, 20 meters, and 15 meters
D. 80 meters, 20 meters, 15 meters, and 10 meters

These restricted segments are designed to ensure that higher-class license holders have their own space on the bands.

Frequency Privileges for General Class Licensees

General class licensees have access to a large portion of the HF bands, but they are not permitted to operate as control operators in certain frequency ranges. These restricted segments are reserved for operators with higher-level licenses. Knowing which frequencies are off-limits is important to ensure compliance with FCC regulations.

General class licensees are prohibited from operating as control operators on frequencies between 7.125 MHz and 7.175 MHz. This range falls within the 40-meter band, but only Amateur Extra license holders are allowed to operate in this portion of the band. General license holders must stay outside this segment when operating as control operators to avoid operating beyond their legal privileges.

On which of the following frequencies are General class licensees prohibited from operating as control operator? (G1A05)

A. 7.125 MHz to 7.175 MHz
B. 28.000 MHz to 28.025 MHz
C. 21.275 MHz to 21.300 MHz
D. All of the above

Phone Restrictions

Some bands are restricted to certain modes of operation, meaning that specific activities, like voice (phone) transmissions, may be prohibited. These restrictions help manage spectrum use and avoid interference, ensuring that different modes of communication can coexist efficiently. Understanding which bands allow or restrict certain types of operation is needed to comply with FCC regulations.

The 30-meter band is one of the HF bands where phone operation (voice communication) is prohibited. This band is reserved for digital modes and CW (Morse code) transmissions. The prohibition of phone operation on this band helps avoid congestion and interference, especially since 30 meters is often used for long-distance digital communication under challenging propagation conditions.

On which of the following bands is phone operation prohibited? (G1A02)

A. 160 meters
B. 30 meters
C. 17 meters
D. 12 meters

The 30-meter band is a "digital-only" band where operators can use CW and data modes but not voice communication. This restriction ensures efficient band use for specific communication modes that benefit from quieter, less crowded conditions.

Image Restrictions

In amateur radio, different frequency bands have specific regulations regarding which modes and types of communication are allowed. One of these modes is image transmission, which includes sending pictures or visual data, typically through modes like SSTV (Slow Scan Television). Some bands, however, have restrictions on image transmission to maintain their designated use for other communication types, such as digital modes and CW (Morse code). Understanding where image transmission is prohibited helps operators stay within legal guidelines and use the spectrum efficiently.

Image transmission is prohibited on the 30-meter band. This band is designated primarily for digital modes and CW, and voice and image transmissions are not allowed. The prohibition on image transmission helps keep the band clear for its intended uses, particularly for long-distance CW and data communications, which require a less congested environment to function optimally.

On which of the following bands is image transmission prohibited? (G1A03)

A. 160 meters
B. 30 meters
C. 20 meters
D. 12 meters

Again, the exam is asking about the 30-meter band. This band is unknown as a "quiet" band reserved for Morse code and data communication, where both phone and image transmissions are prohibited to avoid interference with the more narrow-bandwidth modes that rely on this space.

Chapter Summary

This chapter has explored the diverse frequency privileges and band plans available to General class licensees, highlighting the opportunities and responsibilities that come with expanded access to the amateur radio spectrum. By understanding ITU regions, channelized bands, and specific regulations for bands like 60 meters, you gain insight into the broader scope of your operating privileges. We've also examined critical aspects like restrictions, modulation types, and the shared use of certain frequencies.

Band Plan

Chapter 26

Adhering to FCC Rules

COMPLIANCE WITH FCC RULES is a legal obligation and a fundamental aspect of being a responsible ham radio operator. Following these regulations ensures that every operator can use the spectrum fairly. The radio spectrum is a limited resource, and equitable access is crucial for all operators to enjoy their hobby without causing or experiencing interference. Adhering to these rules contributes to a well-organized and respectful environment where everyone can communicate effectively.

A lot of the basic and universal rules were covered on the Technician's exam.

As a General Class licensee, you must be familiar with specific FCC rules that apply to your operating privileges. Frequency allocations and band limits are defined to ensure that operators use designated frequencies responsibly. General Class operators can access a wider range of frequencies, including portions of the HF bands, which allow for long-distance communications. Knowing the exact frequencies and band segments you are authorized to use is necessary to avoid interfering with other services.

Power limits and restrictions vary across different bands. While the General Class license allows for increased power output, the best practice is to use the minimum power necessary to maintain effective communication. This practice helps reduce interference with other operators and ensures efficient use of the spectrum. Familiarize yourself with the power limits for each band and adhere to them strictly.

The FCC also clearly defines authorized modes of operation and emission types. As a General Class operator, you can use various modes, including CW, SSB, AM, FM, and digital modes like PSK31 and FT8. Each mode has specific emission types and bandwidth requirements that must be followed. Understanding these parameters ensures that your transmissions are within legal limits and do not cause undue interference.

Operating Rules

Operators must follow specific regulations outlined in Part 97 of the FCC rules and adhere to good engineering and amateur practice principles. This concept refers to conducting operations in a way that ensures the best technical standards are met, promotes efficient spectrum use, and avoids interference with other users. These guidelines extend beyond the letter of the law, promoting responsible behavior among amateur operators.

The FCC determines what constitutes good engineering and good amateur practice for amateur radio operators. While many aspects of amateur radio operations are clearly defined in Part 97, these rules do not cover some situations. In such cases, the FCC expects operators to follow established best practices in their technical operations and conduct. This includes ensuring that their transmissions are clean, clear of interference, and within the proper power limits, even when specific details are not laid out in the regulations.

Who or what determines "good engineering and good amateur practice," as applied to the operation of an amateur station in all respects not covered by the Part 97 rules? (G1B11)

A. The FCC
B. The control operator
C. The IEEE
D. The ITU

Permitted Transmissions

Amateur radio operators are generally restricted from retransmitting third-party communications. Still, there are exceptions for specific types of valuable public information. One of these exceptions is the occasional retransmission of weather and propagation forecasts from authorized US government stations. This rule allows amateur radio operators to share important information that benefits the wider amateur community, particularly when it comes to safety or enhancing radio communication effectiveness.

All amateur stations are permitted to occasionally retransmit weather and propagation forecast information from US government stations. This exception allows hams to share important forecasts, such as severe weather alerts or propagation reports, that can affect communication or public safety. While amateur operators are generally not allowed to retransmit content from other services, weather, and propagation information from official government sources is an important exception.

Which of the following transmissions is permitted for all amateur stations? (G1B04)

A. Unidentified transmissions of less than 10 seconds duration for test purposes only
B. Automatic retransmission of other amateur signals by any amateur station
C. Occasional retransmission of weather and propagation forecast information from US government stations
D. Encrypted messages, if not intended to facilitate a criminal act

One-Way Transmissions

Hopefully, you remember from the Technician's exam that amateur radio transmissions are generally intended to be two-way communications between operators. However, there are certain exceptions where one-way transmissions are allowed. One key exception is for transmissions that assist with learning and practicing the International Morse code. These transmissions can be valuable for operators who are working to improve their Morse code proficiency, which remains a popular and important mode in amateur radio.

One-way transmissions that are specifically allowed under FCC rules include those made to assist with learning the International Morse code. This means that operators can transmit Morse code practice messages or educational content to help other hams develop their skills without needing a return signal. These transmissions support the amateur radio community by fostering skills in CW (continuous wave) operation, which is still widely used today.

Which of the following one-way transmissions are permitted? (G1B05)

A. Unidentified test transmissions of less than 10 seconds in duration
B. Transmissions to assist with learning the International Morse code
C. Regular transmissions offering equipment for sale, if intended for amateur radio use
D. All these choices are correct

It's one of the few exceptions to the general rule that amateur radio transmissions should be two-way.

Harmful Interference

Radio operators are responsible for ensuring that their transmissions do not cause harmful interference to other users or services. The FCC has specific guidelines to ensure that operators take extra precautions when transmitting under certain conditions, such as near FCC monitoring stations, on bands where the amateur service is secondary, or when using spread spectrum emissions. These rules protect the shared use of the radio spectrum and ensure cooperation among different services and users.

A licensed amateur radio operator must take specific steps to avoid harmful interference under three key conditions:

1. **Operating within one mile of an FCC Monitoring Station**: These stations monitor the radio spectrum for unauthorized transmissions and interference. To avoid interfering with their work, operators must ensure their signals are clean and well-controlled when operating nearby.

2. **Using a band where the Amateur Service is secondary**: In some frequency bands, amateur radio is designated as a secondary service, meaning it must not interfere with primary users, such as government or commercial services. Operators must be especially careful in these cases, ensuring they don't interfere with primary users.

3. **Transmitting spread spectrum emissions**: Spread spectrum transmissions cover a wider bandwidth and have the potential to interfere with other users. Extra care must be taken to minimize interference, as these transmissions can overlap with other signals.

Which of the following conditions require a licensed amateur radio operator to take specific steps to avoid harmful interference to other users or facilities? (G1E04)

A. When operating within one mile of an FCC Monitoring Station
B. When using a band where the Amateur Service is secondary
C. When a station is transmitting spread spectrum emissions
D. All these choices are correct

Consider these situations as special cases where additional care is required to avoid interference, protecting both the amateur service and other radio spectrum users.

Transmitted Communications

Operators often use abbreviations and procedural signals (such as Q-codes or prosigns) to simplify and speed up their transmissions. These shorthand signals help reduce the time spent transmitting while conveying commonly understood meanings. However, the use of such signals is regulated to ensure that the intended meaning of a message remains clear and understandable to other operators.

Abbreviations and procedural signals are allowed in amateur radio, but they must not obscure the message's meaning. This means that while using shorthand is helpful for efficient communication, it is important that these signals do not make the message difficult to understand. For instance, Q-codes like "QSL" (acknowledge receipt) or "QRZ" (who is calling me?) are widely accepted, but using them in a way that confuses the listener or hides the message's intent is not permitted.

What are the restrictions on the use of abbreviations or procedural signals in the amateur service? (G1B07)

A. Only "Q" signals are permitted
B. They may be used if they do not obscure the meaning of a message
C. They are not permitted
D. They are limited to those expressly listed in Part 97 of the FCC rules

Remember back to the Technician's exam when we learned that only space station or radio-controlled craft communications are allowed to have messages that obscure their meaning? So, always ensure you don't cloud the meaning of your message.

International Communications

Amateur radio provides operators the exciting opportunity to communicate with other hams worldwide. However, these international communications are subject to specific rules and agreements governed by the FCC and the International Telecommunication Union (ITU). While most countries support amateur radio communication, there are a few exceptions where certain governments have objected to allowing such communication across their borders. Understanding these rules is essential for General class operators who want to make contacts beyond the U.S.

Communicating with amateur stations in countries outside the areas administered by the FCC is permissible as long as those countries have not notified the ITU that they object to such communications. In other words, most international contacts are allowed unless a country explicitly requests to restrict amateur radio communications with other nations. This rule ensures that international amateur radio contacts comply with global agreements and respect the laws of other countries.

When is it permissible to communicate with amateur stations in countries outside the areas administered by the Federal Communications Commission? (G1B08)

A. Only when the foreign country has a formal third-party agreement filed with the FCC
B. When the contact is with amateurs in any country except those whose administrations have notified the ITU that they object to such communications
C. Only when the contact is with amateurs licensed by a country which is a member of the United Nations, or by a territory possessed by such a country
D. Only when the contact is with amateurs licensed by a country which is a member of the International Amateur Radio Union, or by a territory possessed by such a country

To remember this for the exam, think of it as a general rule: you can talk to hams worldwide, except in countries whose governments have told the ITU that they do not want such communications—as of this writing, only North Korea and Yemen. This keeps international amateur radio contacts within legal boundaries and respects international agreements.

International Remote Control

With technological advancements, operating a US amateur radio station remotely has become increasingly common. However, when an operator controls a US-based station from outside the country, certain licensing requirements apply to ensure compliance with FCC rules. These regulations ensure that remote operations are conducted by individuals authorized under US law, even when controlling the station from international locations.

When operating a US station by remote control from outside the country, the operator must hold a valid US operator/primary station license. This means that even though you may be physically located in another country while remotely controlling a US-based station, you must still hold the appropriate FCC-issued license to operate legally. The rule is in place to ensure that all remote operations fall under US jurisdiction and are held to the same standards as any local operation within the United States.

When operating a US station by remote control from outside the country, what license is required of the control operator? (G1D05)

A. A US operator/primary station license
B. Only an appropriate US operator/primary license and a special remote station permit from the FCC
C. Only a license from the foreign country, as long as the call sign includes identification of portable operation in the US
D. A license from the foreign country and a special remote station permit from the FCC

However, the regulations for controlling a station remotely depend on the country where the remote station is located. This ensures that operations comply with the laws of the station's location, regardless of where the control operator is physically present.

For example, when operating a station in South America remotely from the US, the operator must follow the regulations of the remote station's country. This means that the rules governing the station's operation are set by the country in which the station is physically located, not by the FCC or US regulations. The operator must ensure they are familiar with and comply with that specific country's licensing and operational standards, even if controlling the station from a location outside its borders, such as from the United States.

When operating a station in South America by remote control over the internet from the US, what regulations apply? (G1D12)

A. Those of both the remote station's country and the FCC
B. Those of the remote station's country and the FCC's third-party regulations
C. Only those of the remote station's country
D. Only those of the FCC

Third-Party Transmissions

Third-party communication involves relaying a message between two amateur radio operators with a third, non-licensed person participating in the exchange. The FCC sets rules to ensure that third-party communication follows amateur radio guidelines, especially regarding licensing status and international agreements. These rules aim to maintain the integrity and proper use of amateur radio frequencies while allowing for important personal or emergency-related communications.

A third party is disqualified from participating in an amateur radio message if their amateur license has been revoked and not reinstated. This rule ensures that only properly authorized individuals can participate in amateur radio communication, maintaining compliance with FCC standards.

Which of the following would disqualify a third party from participating in sending a message via an amateur station? (G1E01)

A. The third party's amateur license has been revoked and not reinstated
B. The third party is not a US citizen
C. The third party is speaking in a language other than English
D. All these choices are correct

Messages sent to a third party in a country with which the U.S. has a Third-Party Agreement must be limited to amateur radio-related topics, personal remarks, or communications related to emergencies or disaster relief. This restriction ensures that the content aligns with the purpose of amateur radio and avoids commercial or inappropriate use of the frequencies.

What are the restrictions on messages sent to a third party in a country with which there is a Third-Party Agreement? (G1E05)

A. They must relate to emergencies or disaster relief
B. They must be for other licensed amateurs
C. They must relate to amateur radio, or remarks of a personal character, or messages relating to emergencies or disaster relief
D. The message must be limited to no longer than 1 minute in duration, and the name of the third party must be recorded in the station log

---

Finally, third-party messages can be transmitted via remote control under any circumstances in which the FCC permits them. This includes situations that follow the rules for third-party communication, whether the station is operated remotely or in person.

When may third-party messages be transmitted via remote control? (G1E12)

A. Under any circumstances in which third-party messages are permitted by FCC rules
B. Under no circumstances except for emergencies
C. Only when the message is intended for licensed radio amateurs
D. Only when the message is intended for third parties in areas where licensing is controlled by the FCC

By adhering to these rules, amateur radio operators ensure that third-party communication is conducted properly, maintaining the purpose and spirit of amateur radio.

Cross-Band Operations

Repeaters are commonly used to extend the range of communication by receiving a signal on one frequency and retransmitting it on another. Cross-band repeaters, such as those operating between the 10-meter and 2-meter bands, allow signals to be received on one band and transmitted on another. However, the operator's license class controlling the repeater affects what frequencies can be legally used for retransmission, as FCC rules are designed to ensure that higher-class licensees are responsible for operations on bands that require more advanced privileges.

A 10-meter repeater can only retransmit a 2-meter signal from a Technician class control operator if the control operator of the 10-meter repeater holds at least a General class license. This is because transmitting on the 10-meter band requires General class privileges or higher. In contrast, Technician class operators have limited privileges on HF bands. The General or higher class repeater control operator ensures compliance with FCC regulations by taking responsibility for the proper operation on the 10-meter band.

When may a 10-meter repeater retransmit the 2-meter signal from a station that has a Technician class control operator? (G1E02)

A. Under no circumstances
B. Only if the station on 10-meters is operating under a Special Temporary Authorization allowing such retransmission
C. Only during an FCC-declared general state of communications emergency
D. Only if the 10-meter repeater control operator holds at least a General class license

The repeater operator is responsible for ensuring legal operation on the higher frequency.

Digital Station Control

Digital stations are often operated under automatic control, which means the station can function without direct operator intervention. However, strict rules apply when conducting digital communications outside designated automatic control band segments, especially to avoid interference and ensure proper control. Operators must maintain local or remote control to ensure compliance with FCC regulations when operating outside these segments.

To conduct communications with a digital station under automatic control outside the designated automatic control segments, the station initiating the contact must be under local or remote control. This ensures an operator can intervene and take responsibility for the communication, avoiding unintended interference. Automatic control is generally allowed only within certain segments of the band, but outside of these areas, active control by the operator is required.

What is required to conduct communications with a digital station operating under automatic control outside the automatic control band segments? (G1E03)

A. The station initiating the contact must be under local or remote control
B. The interrogating transmission must be made by another automatically controlled station
C. No third-party traffic may be transmitted
D. The control operator of the interrogating station must hold an Amateur Extra class license

Automatically controlled stations that transmit RTTY (Radio Teletype) or data emissions can communicate with other automatically controlled digital stations on 6-meter or shorter wavelength bands and in limited segments of some HF bands. These specific allocations are designed to minimize interference while still allowing for automatic communication on certain frequencies.

On what bands may automatically controlled stations transmitting RTTY or data emissions communicate with other automatically controlled digital stations? (G1E11)

A. On any band segment where digital operation is permitted
B. Anywhere in the non-phone segments of the 10-meter or shorter wavelength bands
C. Only in the non-phone Extra Class segments of the bands
D. Anywhere in the 6-meter or shorter wavelength bands, and in limited segments of some of the HF bands

To summarize both questions, it is important to follow FCC rules carefully when operating digital stations under automatic control. Outside of the designated segments, operators must have local or remote control to ensure proper oversight. Automatically controlled digital stations are restricted to certain bands, including the 6-meter and shorter wavelength bands and certain parts of the HF bands for RTTY and data emissions. These rules are in place to ensure the responsible use of automatic control in the amateur radio service.

New Digital Protocols

Digital modes have expanded rapidly, and operators often experiment with new digital communication protocols. However, these new protocols must comply with FCC rules to ensure transparency and interoperability across the amateur community. Before putting a new digital protocol on the air, certain steps must be taken to ensure its technical characteristics are available for public scrutiny and avoid interference or confusion with other established modes.

Before using a new digital protocol on the air, operators must publicly document its technical characteristics. This rule ensures that other operators and regulators can understand how the protocol works and verify that it complies with FCC regulations.

Public documentation helps maintain transparency, allowing the amateur community to use and experiment with the new protocol without causing interference or violating rules.

What must be done before using a new digital protocol on the air? (G1C07)

A. Type-certify equipment to FCC standards
B. Obtain an experimental license from the FCC
C. Publicly document the technical characteristics of the protocol
D. Submit a rule-making proposal to the FCC describing the codes and methods of the technique

---

Make sure everyone knows how the new digital protocol operates before it's used on the airwaves. By documenting the protocol publicly, the amateur radio community ensures responsible and compliant experimentation and innovation.

Power Regulations

The FCC regulates the maximum power amateur radio stations can transmit to ensure that operators use the radio spectrum responsibly and avoid interference with other services. The measurement used to regulate this maximum power is called PEP (Peak Envelope Power), which refers to the highest level of power output during a transmission. Understanding this measurement helps amateur operators comply with FCC power limits and maintain good amateur practice.

The FCC specifies that maximum power limits are measured in terms of PEP output from the transmitter. PEP represents the highest instantaneous power level in the radio signal's envelope, typically measured during voice or data transmission peaks. The FCC restricts operators to a maximum of 1500 watts PEP for most amateur bands. This measurement ensures that the transmitted signal remains within safe and legal limits, preventing interference with other users on the band.

What measurement is specified by FCC rules that regulate maximum power? (G1C11)

A. RMS output from the transmitter
B. RMS input to the antenna
C. PEP input to the antenna
D. PEP output from the transmitter

The FCC imposes specific power regulations across different bands for General class licensees. Here's an overview of the key power regulations:

1. **Maximum Power Limit**: General class licensees are permitted to transmit with up to **1500 watts PEP** (Peak Envelope Power) on most amateur HF, VHF, and UHF bands. The FCC allows the highest power for amateur radio operations on most frequencies.

2. **Exceptions and Lower Power Limits**:

   ◦ **30-meter band (10.100 to 10.150 MHz)**: The maximum power limit is **200 watts PEP**. This band is shared with other services, and the lower power limit helps minimize interference.

   ◦ **60-meter band (5 channels between 5.3305 and 5.3715 MHz)**: The maximum power is **100 watts PEP**, and transmissions must not exceed the equivalent of 100 watts PEP relative to an isotropic radiator (EIRP).

   ◦ **Certain VHF/UHF bands**: Power limits on the 2200-meter band (135.7-137.8 kHz) and 630-meter band (472-479 kHz) are typically lower, restricted to **1 watt EIRP** or less.

3. **Restrictions on Specific Modes**: Certain digital and weak signal modes may also have specific power limitations to prevent interference, but these are rare. In general, using the **minimum power necessary** to maintain commu-

nication is recommended.

With that overview, let's review some exam questions.

What is the maximum transmitter power an amateur station may use on 10.140 MHz? (G1C01)

A. 200 watts PEP output
B. 1000 watts PEP output
C. 1500 watts PEP output
D. 2000 watts PEP output

The 30-meter band is a unique part of the HF spectrum, shared with other radio services, which necessitates stricter regulations on power limits. As a result, the FCC imposes lower power restrictions on this band to avoid interference with other users. While General class licensees have the privilege to operate on this band, they must adhere to these power limitations.

On 10.140 MHz, which is part of the 30-meter band, amateur stations are restricted to a maximum transmitter power of 200 watts PEP (Peak Envelope Power). The 30-meter band has stricter regulations than other HF bands because it is shared with other services, including international broadcasts and fixed services. This power limit ensures that amateur radio operators do not interfere with these other critical services while still allowing effective communication within the amateur community.

What is the maximum transmitter power an amateur station may use on the 12-meter band? (G1C02)

A. 50 watts PEP output
B. 200 watts PEP output
C. 1500 watts PEP output
D. An effective radiated power equivalent to 100 watts from a half-wave dipole

---

The 12-meter band is part of the HF spectrum allocated for amateur radio use. It offers excellent opportunities for long-distance communication, particularly during periods of high solar activity.

You will notice that the exceptions above don't mention this band. Therefore, for the 12-meter band, an amateur station's maximum power is 1500 watts PEP (Peak Envelope Power). This is the standard maximum power limit for most HF bands under the FCC regulations, and it allows operators to transmit with sufficient power for long-distance communication. This power level is particularly useful on the 12-meter band during times of favorable propagation, such as when the solar cycle is active, enabling effective global communication.

What is the limit for transmitter power on the 28 MHz band for a General Class control operator? (G1C05)

A. 100 watts PEP output
B. 1000 watts PEP output
C. 1500 watts PEP output
D. 2000 watts PEP output

Again, we don't see this frequency called out in the exceptions. So, on the 28 MHz band (also known as the 10-meter band), General class control operators are allowed to use up to 1500 watts of PEP (Peak Envelope Power) output. This power limit enables operators to maximize their signal strength, which is particularly beneficial when conditions on the band are less favorable. The high power limit on this band makes it easier for hams to achieve long-distance contacts, especially during periods of sunspot activity, when the band can support global communications.

What is the limit for transmitter power on the 1.8 MHz band? (G1C06)

A. 200 watts PEP output
B. 1000 watts PEP output
C. 1200 watts PEP output
D. 1500 watts PEP output

The 160-meter band, centered around 1.8 MHz, is often referred to as the "top band" and is the lowest-frequency HF band available to amateur radio operators. This band offers a unique challenge for long-distance communication, especially at night when atmospheric conditions allow for better propagation. The FCC permits high power levels on this band to help overcome the challenges of lower-frequency communication.

Have you found the pattern yet?

If not called out as an exception, then on the 1.8 MHz band, the maximum allowable transmitter power for a General class operator is 1500 watts PEP). This high power limit is particularly important for 160 meters, as lower frequencies tend to experience more signal attenuation and require more power to maintain effective communication. The combination of the high power limit and the unique propagation characteristics of this band provides excellent opportunities for long-distance communication, especially during nighttime hours when the band tends to open up.

And our last question on power restrictions.

What is the maximum power limit on the 60-meter band? (G1C09)

A. 1500 watts PEP
B. 10 watts RMS
C. ERP of 100 watts PEP with respect to a dipole
D. ERP of 100 watts PEP with respect to an isotropic antenna

First, you see, it is an exception. So, special rules apply.

The 60-meter band is unique in amateur radio as it operates on five specific channels rather than across a frequency range, and it is shared with other users, such as federal agencies. Because of this shared usage, the FCC imposes stricter power limits and other regulations to ensure that amateurs do not interfere with other services. Power on the 60-meter band is measured in terms of Effective Radiated Power (ERP), which considers the antenna's gain relative to a dipole.

On the 60-meter band, the maximum allowable power is 100 watts PEP (Peak Envelope Power) ERP, with respect to a dipole antenna. This means that the transmitted power must not exceed the equivalent of 100 watts as a standard dipole antenna would radiate it. This restriction is important because the 60-meter band is shared with government and non-amateur services, and limiting power helps to prevent interference with these primary users.

Spread Spectrum Transmissions

Spread spectrum is a transmission technique used in radio that spreads the signal over a wider bandwidth than necessary. It provides benefits such as resistance to interference, secure communication, and reduced potential for signal interception. While spread spectrum is often used in commercial and military applications, it is also permitted in the amateur radio service under certain restrictions, particularly regarding power limits.

The maximum Peak Envelope Power (PEP) output allowed for spread spectrum transmissions in amateur radio is 10 watts. The FCC sets this lower power limit to minimize the potential for interference with other users in the band, given that spread spectrum signals occupy a wider bandwidth. The 10-watt limit ensures that spread spectrum transmissions remain low-power while still taking advantage of their interference-resistant properties.

What is the maximum PEP output allowed for spread spectrum transmissions? (G1E08)

A. 100 milliwatts
B. 10 watts
C. 100 watts
D. 1500 watts

To remember this for the exam, associate spread spectrum with its unique ability to handle interference and the requirement to operate within a 10-watt PEP power limit to avoid disrupting shared frequencies.

Antenna Structures

Height Regulations

Antennas are crucial in optimizing communication, and their height can significantly impact performance. However, there are federal regulations in place to ensure that antenna structures do not interfere with air traffic. For antennas not located near public-use airports, operators must still be mindful of height limits set by the FAA (Federal Aviation Administration) and FCC (Federal Communications Commission). These regulations help ensure safety in air navigation while allowing amateur operators to enjoy the benefits of elevated antenna installations.

The maximum height for an antenna structure without requiring notification to the FAA and registration with the FCC is 200 feet above ground level, provided it is not near a public-use airport. This limit is set to avoid potential hazards to air navigation. Any structure exceeding this height must be reported to the FAA and registered with the FCC to ensure that it does not pose a risk to aircraft. Antenna height can greatly improve signal reach, but safety regulations must always be followed.

What is the maximum height above ground for an antenna structure not near a public-use airport without requiring notification to the FAA and registration with the FCC? (G1B01)

A. 50 feet
B. 100 feet
C. 200 feet
D. 250 feet

Two hundred feet is the safe height limit for antennas in areas not close to public airports. Anything taller requires official clearance to ensure that it does not interfere with air traffic.

Local Restrictions

While HAM radio is governed primarily by federal regulations, state and local governments also have some authority to regulate antenna structures. However, such regulations must ensure that amateur radio operators are reasonably accommodated in their communication ability. This balance allows local authorities to address safety, aesthetic, or zoning concerns while ensuring these regulations do not overly burden or prevent amateur radio operations.

State and local governments are permitted to regulate amateur radio antenna structures under certain conditions, but they must ensure that the regulations allow for amateur service communications to be reasonably accommodated. This means that regulations cannot unnecessarily restrict an operator's ability to set up and use antennas. The minimum regulations must also be necessary to serve a legitimate purpose, such as safety or zoning requirements. For example, a local government might limit antenna heights for safety reasons, but they cannot outright ban antennas without a valid reason.

Under what conditions are state and local governments permitted to regulate amateur radio antenna structures? (G1B06)

A. Under no circumstances, FCC rules take priority
B. At any time and to any extent necessary to accomplish a legitimate purpose of the state or local entity, provided that proper filings are made with the FCC
C. Only when such structures exceed 50 feet in height and are clearly visible 1,000 feet from the structure
D. Amateur Service communications must be reasonably accommodated, and regulations must constitute the minimum practical to accommodate a legitimate purpose of the state or local entity

---

To remember this for the exam, consider the regulations as needing to strike a balance: They must protect legitimate local interests while ensuring that hams can operate their stations effectively. Regulations must be the least restrictive necessary to achieve their purpose.

Beacon Stations

A beacon station is an amateur radio station that continuously transmits a signal on a set frequency to provide information about propagation conditions or serve as a reference signal for other operators. Beacon stations typically transmit a simple, repetitive signal, often including the station's call sign and other identifying information, such as location or operating frequency.

According to FCC rules, a beacon station's primary purpose is to observe propagation and reception. This means that beacon stations are set up to help operators monitor how radio waves travel across different regions, which bands are open for communication, and how conditions change over time. Operators can use the information provided by beacon stations to decide which frequencies might be optimal for their communication.

Which of the following is a purpose of a beacon station as identified in the FCC rules? (G1B03)

A. Observation of propagation and reception
B. Automatic identification of repeaters
C. Transmission of bulletins of general interest to amateur radio licensees
D. All these choices are correct

Beacon stations are valuable for determining how well radio waves are propagating over long distances at a given time. By listening for beacon signals on different frequencies, other operators can assess the current state of the ionosphere and make informed decisions about which frequencies might be optimal for communication.

Key Points:

- **Propagation testing**: Beacons help operators check signal propagation conditions.

- **Repetitive signal**: Beacons usually send a simple, repeating transmission with their call sign.

- **Frequency reference**: Beacons provide a stable signal that operators can tune to for testing equipment or making adjustments.

Beacon Station Restrictions

Since beacon stations are continuously active, specific regulations ensure they do not interfere with other operations or unnecessarily congest the bands.

One important regulation for beacon stations is that no more than one beacon station may transmit on the same band from the same station location. This rule helps prevent multiple beacons from creating interference or congestion on a single frequency band. By limiting a location to one beacon per band, the FCC ensures that the spectrum is used efficiently and that operators can rely on clean, clear signals from each beacon station.

With which of the following conditions must beacon stations comply? (G1B02)

A. No more than one beacon station may transmit in the same band from the same station location
B. The frequency must be coordinated with the National Beacon Organization
C. The frequency must be posted on the internet or published in a national periodical
D. All these choices are correct

Multiple beacons on the same band from the same place could cause confusion and interfere with the station's purpose of providing propagation information.

Automatically Controlled Beacon Stations

In the HF bands, beacons can either be manually controlled or automatically controlled, meaning they transmit without an operator needing to be physically present. However, automatic control of beacons is only permitted on specific frequency ranges to prevent interference with other communications.

Automatically controlled beacon stations are permitted on HF frequencies ranging from 28.20 MHz to 28.30 MHz, which falls within the 10-meter band. This 100 kHz segment is reserved for beacons to transmit signals continuously, helping operators monitor propagation conditions on this band. Automatic control allows beacons to transmit without the need for constant operator input, ensuring they remain on air for extended periods.

On what HF frequencies are automatically controlled beacons permitted? (G1B09)

A. On any frequency if power is less than 1 watt
B. On any frequency if transmissions are in Morse code
C. 21.08 MHz to 21.09 MHz
D. 28.20 MHz to 28.30 MHz

Remember this for the exam: 28.20 MHz to 28.30 MHz as the dedicated beacon segment on the 10-meter band, where beacons can run automatically, providing useful propagation data for operators tuning in.

Power Limits

Beacon stations must operate under specific power limitations to prevent interference with other spectrum users and ensure that they can coexist efficiently with other amateur services.

The power limit for beacon stations is 100 watts of PEP (Peak Envelope Power) output. This means that beacons are restricted to a maximum of 100 watts of output power during their transmissions. This limitation helps ensure that beacons do not overpower other stations on the band and allows for efficient use of the spectrum, particularly on shared frequency ranges.

> **What is the power limit for beacon stations? (G1B10)**
>
> A. 10 watts PEP output
> B. 20 watts PEP output
> **C. 100 watts PEP output**
> D. 200 watts PEP output

Transmission Restrictions

Amateur operators should avoid transmitting on 14.100 MHz, 18.110 MHz, 21.150 MHz, 24.930 MHz, and 28.200 MHz because a system of propagation beacon stations uses these frequencies. These beacons, part of an international network, continuously transmit signals that help operators assess propagation conditions across various parts of the radio spectrum. By listening to these beacons, amateur operators can determine which bands are open for long-distance communication. Transmitting on these frequencies would interfere with the beacons' signals, disrupting their use for propagation analysis.

> **Why should an amateur operator normally avoid transmitting on 14.100, 18.110, 21.150, 24.930 and 28.200 MHz? (G1E10)**
>
> **A. A system of propagation beacon stations operates on those frequencies**
> B. A system of automatic digital stations operates on those frequencies
> C. These frequencies are set aside for emergency operations
> D. These frequencies are set aside for bulletins from the FCC

Avoiding transmission on these frequencies ensures that these valuable tools remain functional and interference-free for all amateur operators.

Chapter Summary

To stay informed about FCC rules and regulations, utilize various resources available to ham radio operators. The official FCC website offers comprehensive information on current regulations, updates, and bulletins. Subscribing to ham radio newsletters and magazines can provide regular updates and insights on regulatory changes and best practices. Joining ham radio organizations like the American Radio Relay League (ARRL) offers access to valuable resources, training, and community support.

Staying updated on FCC rules is necessary for maintaining compliance and ensuring a positive operating experience.

Chapter 27

Rules for the General Exam

AFTER PASSING A LICENSING exam, amateur operators receive a Certificate of Successful Completion of Examination (CSCE). This certificate grants temporary operating privileges for the new license class until the official FCC upgrade is processed. For Technician class operators, this means they can start using General class privileges while waiting for their license upgrade, expanding the range of frequencies they can operate on.

Suppose you are a Technician class operator with an unexpired CSCE for General class privileges. In that case, you are allowed to operate on any General or Technician class band segment. This means that as soon as you pass the General exam, you can immediately start using both General and Technician frequency allocations, even before your official license is upgraded in the FCC database. However, these privileges are temporary and last for 365 days or until your upgrade is officially processed; after 365 days, your CSCE expires.

On which of the following band segments may you operate if you are a Technician class operator and have an unexpired Certificate of Successful Completion of Examination (CSCE) for General class privileges? (G1D03)

A. Only the Technician band segments until your upgrade is posted in the FCC database
B. Only on the Technician band segments until you have a receipt for the FCC application fee payment
C. On any General or Technician class band segment
D. On any General or Technician class band segment except 30 meters and 60 meters

---

And

How long is a Certificate of Successful Completion of Examination (CSCE) valid for exam element credit? (G1D09)

A. 30 days
B. 180 days
C. 365 days
D. For as long as your current license is valid

Temporary Operating Privileges

When a Technician class licensee passes the General class exam, they are granted temporary operating privileges before the upgrade is processed in the FCC database. These privileges allow the operator to use General class frequency bands. Still, specific identification rules must be followed during this interim period. The requirement to use a special identifier ensures that other operators know the licensee is temporarily operating under General privileges, which are not yet reflected in the official FCC records.

Until the General class upgrade is shown in the FCC database, a Technician class licensee must identify with "AG" (short for Authorized General) after their call sign whenever they are using General class frequency privileges. This is a way of informing others on the air that the operator is operating with General privileges based on a CSCE (Certificate of Successful Completion of Examination), but the upgrade has not yet been processed. When operating on Technician class bands, the "AG" is not required, but it must be used every time the operator transmits on General class frequencies.

Until an upgrade to General class is shown in the FCC database, when must a Technician licensee identify with "AG" after their call sign? (G1D06)

A. Whenever they operate using General class frequency privileges
B. Whenever they operate on any amateur frequency
C. Whenever they operate using Technician frequency privileges
D. A special identifier is not required if their General class license application has been filed with the FCC

It's a way of signaling your new privileges to other operators while waiting for the database update.

Renewing an Expired Amateur License

Amateur radio licenses are valid for 10 years. If a license expires, there is a two-year grace period during which the operator can renew their license without taking another exam. However, reinstating the license becomes more complex once this grace period has passed. Operators who want to regain their license privileges after this period must meet certain requirements, including taking parts of the licensing exam again.

Suppose an operator's General class license has expired and the two-year grace period has passed. In that case, the applicant must provide proof of their expired license and pass the current Element 2 exam (the Technician class exam). Element 2 is the foundational exam required for all amateur operators, and passing it reinstates the individual's privileges, including General class, if they can show proof of holding a previously valid General class license. The Element 2 requirement ensures that the operator is up-to-date on basic rules and regulations, while their previous experience is credited for the more advanced privileges.

What action is required to obtain a new General class license after a previously held license has expired and the two-year grace period has passed? (G1D11)

A. They must have a letter from the FCC showing they once held an amateur or commercial license
B. There are no requirements other than being able to show a copy of the expired license
C. Contact the FCC to have the license reinstated
D. The applicant must show proof of the appropriate expired license grant and pass the current Element 2 exam

---

To remember this for the exam, think of it as a two-step process. After the grace period, you must pass the Technician class exam (Element 2) and show proof of your expired General license to regain your operating privileges.

Partial Credit for Expired Licenses

Amateur radio operators who once held higher-class licenses, such as General, Advanced, or Amateur Extra, may be eligible to regain their privileges even if their license has expired. The FCC allows former licensees to avoid retaking all licensing exams by allowing them to receive partial credit for their previously held license. This helps operators re-enter the hobby more easily, recognizing their past achievements while ensuring they still meet current requirements.

Anyone who can demonstrate that they once held an FCC-issued General, Advanced, or Amateur Extra class license, which was not revoked, may receive partial credit for the exam elements that were previously passed. This means that while the operator will need to take the Element 2 exam (Technician), they are not required to retake the General or Extra class exams, as their previous license covers those elements. This provision allows former licensees to restore their privileges without starting from scratch, provided they can show proof of their expired license.

Who may receive partial credit for the elements represented by an expired amateur radio license? (G1D01)

A. Any person who can demonstrate that they once held an FCC-issued General, Advanced, or Amateur Extra class license that was not revoked by the FCC
B. Anyone who held an FCC-issued amateur radio license that expired not less than 5 and not more than 15 years ago
C. Any person who previously held an amateur license issued by another country, but only if that country has a current reciprocal licensing agreement with the FCC
D. Only persons who once held an FCC issued Novice, Technician, or Technician Plus license

Think of it as the FCC's way of honoring past licensees' achievements: if your General, Advanced, or Extra license expired but wasn't revoked, you can get credit for the elements you passed, making it easier to get back on the air.

Volunteer Examiners and Examiner Coordinators

The Volunteer Examiner (VE) program is fundamental in administering amateur radio licensing exams. The FCC authorizes qualified amateur operators to serve as VEs, allowing them to administer licensing exams for new and upgrading operators. These VEs are accredited by Volunteer Examiner Coordinators (VECs), organizations recognized by the FCC to manage and coordinate VEs. The VEs ensure that exams are administered fairly and according to FCC regulations and enable the smooth operation of the amateur radio licensing process.

To administer a Technician class license examination, at least three Volunteer Examiners (VEs) of General class or higher must observe and oversee the exam. This ensures the integrity of the examination process, as multiple VEs confirm the accuracy and fairness of the test administration.

Who must observe the administration of a Technician class license examination? (G1D04)

A. At least three Volunteer Examiners of General class or higher
B. At least two Volunteer Examiners of General class or higher
C. At least two Volunteer Examiners of Technician class or higher
D. At least three Volunteer Examiners of Technician class

Volunteer Examiners are accredited by a Volunteer Examiner Coordinator (VEC), which oversees the accreditation process and ensures that VEs are properly qualified to administer exams. There are several VECs across the United States, with the ARRL VEC being the largest.

Volunteer Examiners are accredited by what organization? (G1D07)

A. The Federal Communications Commission
B. The Universal Licensing System
C. A Volunteer Examiner Coordinator
D. The Wireless Telecommunications Bureau

To qualify as a Volunteer Examiner, <u>one must be at least 18 years old</u> and hold an FCC-granted amateur radio license of General class or above. This rule applies to US and <u>non-US citizens, with the latter requiring a valid FCC-issued General class license or higher.</u>

What is the minimum age that one must be to qualify as an accredited Volunteer Examiner? (G1D10)

A. 16 years
B. 18 years
C. 21 years
D. There is no age limit

Which of the following criteria must be met for a non-US citizen to be an accredited Volunteer Examiner? (G1D08)

A. The person must be a resident of the US for a minimum of 5 years
B. The person must hold an FCC-granted amateur radio license of General class or above
C. The person's home citizenship must be in ITU region 2
D. None of these choices is correct; a non-US citizen cannot be a Volunteer Examiner

A VE holding a General class license may administer exams <u>only for the Technician class</u>. Higher-class exams, such as General or Extra, require Volunteer Examiners with a higher level of licensure.

What license examinations may you administer as an accredited Volunteer Examiner holding a General class operator license? (G1D02)

A. General and Technician
B. None, only Amateur Extra class licensees may be accredited
C. Technician only
D. Amateur Extra, General, and Technician

This system ensures that exams are fair, properly monitored, and in compliance with FCC regulations, allowing qualified individuals to enter the amateur radio hobby.

Chapter Summary

This chapter highlights key elements of amateur radio licensing that directly prepare you for the exam. Understanding the privileges granted by the Certificate of Successful Completion of Examination (CSCE) ensures you know what operating rights you gain immediately after passing an exam. Familiarizing yourself with the roles of Volunteer Examiners and the structure of licensing tests provides clarity on the process and requirements. By mastering these rules and procedures, you'll be better equipped to approach the exam with confidence.

Further Expanding Your Ham Experience

AMATEUR RADIO IS A gateway to a world of opportunities, blending technical expertise with real-world applications that go beyond simply operating a station. In this section, we delve into some of the most rewarding and impactful ways to expand your journey in amateur radio. Whether you're stepping up to provide critical communications during emergencies, supporting the integrity of the airwaves through volunteer monitoring, or pushing your skills and equipment to the limit in HF contesting, amateur radio offers avenues for both service and personal growth. By exploring these activities, you'll uncover the true depth of this hobby—one that fosters community, resilience, and innovation, all while keeping the thrill of discovery alive.

Chapter 28

The Responsibility and Excitement of Amateur Radio

AMATEUR RADIO IS MORE than just a hobby—it's a versatile tool that connects people, communities, and even entire regions during critical times. This chapter dives into three essential areas that showcase the depth and impact of amateur radio operations.

First, we'll explore the Radio Amateur Civil Emergency Service (RACES), a vital program that supports disaster relief and emergency communications. Next, we'll examine the Volunteer Monitor Program, a unique initiative that empowers radio operators to help maintain order and compliance on the airwaves. Finally, we'll delve into the dynamic world of HF contesting, where operators test their skills and equipment under fast-paced, competitive conditions.

Together, these topics reveal amateur radio's multifaceted nature and the important roles operators play in public service and personal achievement.

Radio Amateur Civil Emergency Service

The Radio Amateur Civil Emergency Service (RACES) is an important part of amateur radio operations, designed to assist with disaster relief and emergency communications. It is a partnership between amateur radio operators and emergency management agencies to provide communications during times of crisis. RACES operations are subject to specific rules and regulations, especially concerning who may operate a station under these circumstances. This ensures that only qualified and licensed individuals are responsible for critical emergency communications.

In RACES operations, only a person holding an FCC-issued amateur operator license can act as the station's control operator. This means that during disaster relief efforts, the responsibility for controlling the station lies exclusively with licensed amateurs whom the FCC authorizes. The control operator ensures that all transmissions comply with the rules and that communications aid in emergency or relief operations.

Who may be the control operator of an amateur station transmitting in RACES to assist relief operations during a disaster? (G2B09)

A. Only a person holding an FCC-issued amateur operator license
B. Only a RACES net control operator
C. A person holding an FCC-issued amateur operator license or an appropriate government official
D. Any control operator when normal communication systems are operational

---

This restriction guarantees that RACES stations are operated by qualified individuals who understand the importance of proper communication protocols during critical situations, ensuring smooth and effective disaster response efforts.

How often may RACES training drills and tests be routinely conducted without special authorization? (G2B11)

A. No more than 1 hour per month
B. No more than 2 hours per month
C. No more than 1 hour per week
D. No more than 2 hours per week

RACES training drills and tests can be conducted no more than 1 hour per week without special authorization. This limit allows RACES operators to maintain their skills and be ready for emergencies while ensuring that frequencies are not unnecessarily occupied for extended periods. The 1-hour weekly limit is sufficient for routine practice and maintaining operational readiness. Still, special authorization from the FCC would be necessary if more frequent or longer training sessions are required.

This regulation helps to strike a balance between maintaining emergency preparedness and ensuring that amateur frequencies remain available for general use. By adhering to these rules, operators can be ready to assist during real emergencies while respecting the broader amateur radio community's need for shared frequency access.

Volunteer Monitor Program

The Volunteer Monitor Program is an initiative established by the FCC and managed by the ARRL (American Radio Relay League) that empowers amateur radio operators to help maintain compliance with FCC rules and regulations. These volunteers monitor amateur radio transmissions and help identify and address rule violations. The program fosters a sense of self-regulation within the amateur radio community, allowing operators to take an active role in keeping the airwaves free from harmful interference and illegal activity.

Explanation of the Volunteer Monitor Program

The Volunteer Monitor Program consists of amateur radio operators who are formally enlisted to observe and report on airwaves activity, ensuring that transmissions comply with FCC rules. The volunteers are trained and authorized to monitor the bands for violations, such as improper use of frequencies or intentional interference. The goal is to encourage self-regulation among amateur radio operators and foster a culture of compliance.

What is the Volunteer Monitor Program? (G2D01)

A. Amateur volunteers who are formally enlisted to monitor the airwaves for rules violations
B. Amateur volunteers who conduct amateur licensing examinations
C. Amateur volunteers who conduct frequency coordination for amateur VHF repeaters
D. Amateur volunteers who use their station equipment to help civil defense organizations in times of emergency

Which of the following are objectives of the Volunteer Monitor Program? (G2D02)

A. To conduct efficient and orderly amateur licensing examinations
B. To provide emergency and public safety communications
C. To coordinate repeaters for efficient and orderly spectrum usage
D. To encourage amateur radio operators to self-regulate and comply with the rules

The program also includes a technique for locating problem stations. For example, if a continuous carrier holds a repeater open inappropriately, Volunteer Monitors can use direction-finding techniques. <u>By comparing beam headings on the repeater input from different locations, they can triangulate the source of the signal and assist in addressing the issue.</u>

What procedure may be used by Volunteer Monitors to localize a station whose continuous carrier is holding a repeater on in their area? (G2D03)

A. Compare vertical and horizontal signal strengths on the input frequency
B. Compare beam headings on the repeater input from their home locations with that of other Volunteer Monitors
C. Compare signal strengths between the input and output of the repeater
D. All these choices are correct

This collaborative monitoring effort ensures that the amateur radio bands remain a reliable and well-regulated resource for all operators.

Contesting

Ham radio contesting is a fast-paced activity where operators compete to make as many contacts as possible within a set timeframe, testing their skills and endurance. It fosters community and camaraderie while pushing the limits of amateur radio capabilities. Contests vary widely, from DX contests emphasizing long-distance contacts to VHF/UHF contests challenging operators with line-of-sight propagation, CW contests highlighting Morse code skills, and digital mode contests like RTTY or FT8 requiring digital expertise.

Each contest has unique rules for exchanging information, such as call signs, signal reports, or grid locators, with accurate logging essential for scoring. Points are awarded per contact, with multipliers for different regions or bands and bonuses for achievements like working rare stations. Contests typically last hours to weekends, sometimes requiring mandatory breaks for fairness.

Success in contesting often depends on specialized tools, like logging software (e.g., N1MM Logger+) to track contacts and calculate scores. Advanced transceivers with noise reduction, high-gain antennas like Yagis, and accessories like headsets and foot switches can improve performance. Operators must always follow FCC rules, including proper station identification, to ensure legal and transparent operation.

<u>Operators must identify their station according to normal FCC regulations when participating in a HF frequency contest.</u> This means that despite the rapid pace of contest exchanges, you must still announce your call sign at the required intervals, just as you would during any other communication. This rule helps prevent confusion on the bands and ensures that all operators follow the appropriate identification protocols.

Which of the following is required when participating in a contest on HF frequencies? (G2D09)

A. Submit a log to the contest sponsor
B. Send a QSL card to the stations worked, or QSL via Logbook of The World
C. Identify your station according to normal FCC regulations
D. All these choices are correct

<u>Additionally, many operators use an azimuthal projection map during contests or DXing to help determine the true bearings and distances from their location to other stations.</u> These maps are centered on a specific point, usually the operator's station, and display the correct direction and distance to any location on Earth. This tool is particularly useful for pointing directional antennas to improve signal strength during long-distance communications.

Which of the following describes an azimuthal projection map? (G2D04)

A. A map that shows accurate land masses
B. A map that shows true bearings and distances from a specific location
C. A map that shows the angle at which an amateur satellite crosses the equator
D. A map that shows the number of degrees longitude that an amateur satellite appears to move westward at the equator with each orbit

By combining the awareness of FCC identification rules with practical tools like azimuthal projection maps, contest participants can operate more efficiently while adhering to legal requirements.

Strategies for Winning Contests

Preparation and planning are the bedrock of successful contesting. To maximize your performance, start by researching each contest's specific rules and strategies. Every contest has unique requirements and nuances, so familiarize yourself with the contest's official rules, which can typically be found on the organizing body's website. Understanding the scoring system, exchange requirements, and any special bonuses or multipliers will give you a strategic advantage.

Maximizing your contact rates during a contest requires a combination of effective techniques. Running a frequency, where you hold a frequency and call CQ continuously, can attract a steady stream of contacts. This method is particularly effective when you have a strong signal and a clear frequency. However, choose a frequency that is not too crowded to avoid becoming a target of interference. Alternatively, the search and pounce method involves actively tuning across the band to find and contact other stations. This technique is beneficial when the band is crowded, allowing you to capitalize on every available opportunity.

Chapter Summary

In this chapter, we explored key aspects of amateur radio operations, including the critical roles of RACES in disaster communications, the Volunteer Monitor Program's importance in maintaining compliance, and the excitement and strategy involved in HF contesting. From understanding the unique rules and responsibilities of emergency operations to leveraging tools like azimuthal projection maps during contests, each topic highlights the diverse ways amateur radio operators contribute to public service and personal achievement. Whether preparing for emergencies, fostering self-regulation, or competing in contests, these skills and practices embody the spirit of amateur radio—combining technical expertise, community involvement, and a commitment to operating responsibly and effectively.

Keeping the Hobby Alive & Growing

Keeping the Hobby Alive & Growing

Now that you've taken the step to upgrade to your General Class license, you're in a prime position to explore even more of what ham radio has to offer. But this journey isn't just about you—it's about keeping the spirit of amateur radio alive for future generations.

Would you help someone you've never met, even without getting credit for it? That person might be where you were before you picked up this book—curious about ham radio, eager to expand their knowledge, and looking for guidance. Your honest review could encourage them to take their next step.

At Morse Code Publishing, we aim to make ham radio accessible and exciting for everyone. But to continue growing the hobby and reaching new operators, we need your help. A good book cover catches the eye, but reviews are what truly help others make decisions. By taking just 60 seconds to leave a review, you could:

- Help a beginner discover the joy and appreciate the intricacies of ham radio.

- Help other Technicians prepare for and pass their General Class exam.

Your review costs nothing but could make all the difference in someone else's journey to becoming a licensed operator or upgrading to General.

Thank you for your support! Ham radio grows and thrives when we share our knowledge and enthusiasm. With your help, we can continue to inspire the next generation of operators.

Welcome to the community of General Class operators!

Your biggest fan,

Morse Code Publishing

PS – Helping others succeed is one of the greatest rewards in this hobby. If you believe this book can benefit someone, consider sharing it with them. Let's keep the hobby thriving!

Conclusion

THE JOURNEY THROUGH THIS book was designed with a singular purpose: to provide you, a dedicated ham radio operator, with the knowledge and tools necessary to upgrade from a Technician to a General class license. This guide aims to expand your understanding and capabilities in amateur radio, ensuring you are well-prepared to ace the FCC exam and enjoy the enhanced privileges of the General class license.

From the outset, my vision has been clear—to make complex information accessible and easy to retain. This book has been structured to break down intricate topics into manageable sections, using practical tips and examples to facilitate learning. This approach ensures the material is understandable and memorable, enabling you to apply your newfound knowledge confidently.

The book was organized into four key sections, each addressing crucial aspects of the General class license exam.

We began with the **Science of Radio** section and explored the fascinating principles of radio wave propagation, the types of antennas and their uses, and the intricacies of modulation techniques. This scientific understanding is vital for optimizing your communication range and quality, allowing you to maximize your operating privileges.

Next, in the **Physical Ham Station** section, we covered the technical aspects of setting up and maintaining your ham radio station. From choosing the right equipment to understanding transmitters and receivers and ensuring proper grounding and power supply management, this section provided the practical knowledge necessary to create a functional and efficient station.

In **Operating Procedures**, where you learned the proper techniques for making and answering calls, handling emergency communications, and participating in ham radio nets. These procedures are essential for effective and respectful communication on the airwaves.

Lastly, **Rules & Regulations**, where we explored the role of the FCC, the specific regulations that General class licensees must follow, and the importance of operating legally and ethically. This section laid the foundation for understanding the responsibilities that come with greater privileges.

Key takeaways from this book include a comprehensive understanding of FCC regulations, mastery of operating procedures, the ability to set up and maintain a physical ham station, and a deep knowledge of radio science. These insights will help you pass the General class license exam and enhance your overall ham radio experience.

I encourage you to utilize this guide's study techniques and tips as you prepare for the exam. Practice with the included exams (take all 16!), and remember to manage your time effectively during the test. Your dedication to studying and passion for ham radio will undoubtedly lead you to success.

I wish you the best of luck in your exam and future ham radio endeavors. Keep exploring, learning, and, most importantly, enjoying the incredible world of ham radio.

Jared Johnson (KF0RTU/AG) @ Morse Code Publishing

Exam Question Index

G3 - Radio Wave Propagation

Group G3A

G3A01 25
G3A02 24
G3A03 24
G3A04 28
G3A05 29
G3A06 26
G3A07 23
G3A08 27
G3A09 27
G3A10 23
G3A11 25
G3A12 28
G3A13 28
G3A14 26

Group G3B

G3B01 14
G3B02 19
G3B03 20
G3B04 30
G3B05 15
G3B06 19
G3B07 18
G3B08 18
G3B09 17
G3B10 16
G3B11 21
G3B12 29

Group G3C

G3C01 15
G3C02 17
G3C03 16
G3C04 17
G3C05 15
G3C06 12
G3C07 13
G3C08 13
G3C09 12
G3C10 22
G3C11 15

G4 - Amateur Radio Practices

Group G4A

G4A01 102
G4A02 208
G4A03 100
G4A04 96
G4A05 95
G4A06 150
G4A07 97
G4A08 96
G4A09 103
G4A10 209
G4A11 96
G4A12 103
G4A13 97

Group G4B

G4B01 132
G4B02 133
G4B03 133
G4B04 133
G4B05 132
G4B06 131
G4B07 94
G4B08 95
G4B09 131
G4B10 151
G4B11 154
G4B12 154
G4B13 151

Group G4C

G4C01 221
G4C02 221
G4C03 222
G4C04 223
G4C05 180
G4C06 180
G4C07 181
G4C08 222
G4C09 182
G4C10 223
G4C11 182
G4C12 180

Group G4D

G4D01 104
G4D02 104
G4D03 104
G4D04 100
G4D05 101
G4D06 101
G4D07 101
G4D08 200
G4D09 202
G4D10 200
G4D11 202

Group G4E

G4E01 135
G4E02 136
G4E03 137
G4E04 137
G4E05 134
G4E06 135
G4E07 137
G4E08 168
G4E09 168
G4E10 169
G4E11 168

G5 - Electrical Principles

Group G5A

G5A01 72
G5A02 63
G5A03 64
G5A04 64
G5A05 65
G5A06 65
G5A07 66
G5A08 66
G5A09 64
G5A10 140
G5A11 64
G5A12 67

Group G5B

G5B01 88
G5B02 74
G5B03 83
G5B04 83
G5B05 85
G5B06 110
G5B07 106
G5B08 107
G5B09 108
G5B10 89
G5B11 47
G5B12 108
G5B13 47
G5B14 111

Group G5C

G5C01 77
G5C02 78
G5C03 76
G5C04 75
G5C05 78
G5C06 79
G5C07 80
G5C08 73
G5C09 68
G5C10 74
G5C11 71
G5C12 69
G5C13 72
G5C14 70

References

- *47 CFR Part 97 -- Amateur Radio Service* https://www.ecfr.gov/current/title-47/chapter-I/subchapter-D/part-97

- *General Class Frequency Privileges in Ham Radio* https://www.dummies.com/article/technology/digital-audio-radio/ham-radio/general-class-frequency-privileges-in-ham-radio-164187/

- *1068. Violation of FCC Regulations—47 U.S.C. § 502* https://www.justice.gov/archives/jm/criminal-resource-manual-1068-violation-fcc-regulations-47-usc-502#:~:text=Under%2047%20U.S.C.,on%20which%20a%20violation%20occurs.

- *Frequency Allocations* http://www.arrl.org/frequency-allocations

- *Amateur Radio Service Enforcement - AMAT* https://www.fcc.gov/tags/amateur-radio-service-enforcement-amat

- *Upgrading to a General License* http://www.arrl.org/upgrading-to-a-general-license

- *General Class Frequency Privileges in Ham Radio* https://www.dummies.com/article/technology/digital-audio-radio/ham-radio/general-class-frequency-privileges-in-ham-radio-164187/

- *How to Stop Radio Frequency Interference | PCB Design Blog* https://resources.pcb.cadence.com/blog/2022-how-to-stop-radio-frequency-interference

- *Ethics and Operating Procedures for the Radio Amateur* https://www.arrl.org/files/file/DXCC/Eth-operating-EN-ARRL-CORR-JAN-2011.pdf

- *Choosing a Field Radio: How to find the perfect transceiver ...* https://qrper.com/2022/07/choosing-a-field-radio-how-to-find-the-perfect-transceiver-for-your-outdoor-radio-activities/

- *Improving Receiver Reception: Tips and Tricks* https://www.hamradiostore.co.uk/blog/improving-receiver-reception-tips-and-tricks#:~:text=IF%20Bandwidth%20Adjustment,neighboring%20signals%20and%20background%20noise.

- *Choosing a Power Supply for Your Station * https://www.onallbands.com/choosing-a-power-supply-for-your-station%EF%BB%BF/

- *Ham Radio Tech: Choosing a Battery System for Portable ...* https://www.onallbands.com/ham-radio-tech-choosing-a-battery-system-for-portable-operations/

- *Ham Radio Antennas: Types, Differences, and Pros & Cons* https://strykerradios.com/ham-radios/ham-radio-antenna-types-differences-pros-cons/

- *Amateur Radio Station Grounding and Lightning Protection* https://www.bwcelectronics.com/articles/WP30A190.pdf

- *Basic Amateur Radio - HF Propagation* https://www.arrl.org/files/file/Technology/tis/info/pdf/8312011.pdf

- *Basic Electronic Components - Sierra Circuits* https://www.protoexpress.com/kb/basic-components-overview/#:~:text=Some%20of%20the%20most%20commonly,switches%2C%20ICs%2C%20and%20connectors.

- *How to Read a Schematic - SparkFun Learn* https://learn.sparkfun.com/tutorials/how-to-read-a-schematic/all

- *Loads of Modes - Ham Radio School* https://www.hamradioschool.com/post/loads-of-modes#:~:text=Modulatio n%20Mode%20%E2%80%93%20A%20modulation%20mode,voice%20into%20the%20radio%20transmission.

- *How to Demodulate an FM Waveform | Radio Frequency ...* https://www.allaboutcircuits.com/textbook/radio-fr equency-analysis-design/radio-frequency-demodulation/how-to-demodulate-an-fm-waveform/

- *How to Call and Answer a CQ* https://www.ailunce.com/blog/How-to-Call-and-Answer-a-CQ

- *How to Participate in Ham Radio Nets* https://www.dummies.com/article/technology/digital-audio-radio/ham-r adio/how-to-participate-in-ham-radio-nets-160474/

- *Amateur Radio Digital Communications - SARCNET* https://www.sarcnet.org/amateur-radio-digital-communi cations.html

- *RTTY / FSK Configuration - Fldigi Users Manual* http://www.w1hkj.com/FldigiHelp/rtty_fsk_configuration_p age.htm

- *Amateur Radio Station Grounding* https://w7aia.org/files/operating_files/RCC_Mar_2007_N7ZXP_Amateur _Radio_Station_Grounding.pdf

- *ARRL Handbook: The ARRL Handbook is a comprehensive guide to amateur radio that covers everything from basic electronics to advanced operating techniques.* https://hamradioprep.com/ham-radio-study-guide/

- *Contest Operating Tips, originated by K1AR* https://www.mapability.com/ei8ic/contest/k1ar.php

- *Practical Signal Reports* https://www.hamradioschool.com/post/practical-signal-reports

- *The Ionosphere | Center for Science Education* https://scied.ucar.edu/learning-zone/atmosphere/ionosphere

- *How to Use Meteor Scatter Communications in Ham Radio* https://www.electronics-notes.com/articles/ham_rad io/amateur-propagation/meteor-scatter-burst-communications.php

- *How to Troubleshoot ham radio problems* https://brara.org/BLOG/2018/11/06/new-hams-how-to-troubleshoot -ham-radio-problems-radio-wave-frequency-and-power/

- *Test equipment every amateur radio operator should have* http://forums.radioreference.com/threads/test-equipme nt-every-amateur-radio-operator-should-have.453318/

- *The four pieces of radio test equipment you really need* https://vk3ye.com/gateway/noaug00.htm

- *Ethics and Operating Procedures for the Radio Amateur* https://www.arrl.org/files/file/DXCC/Eth-operating-EN -ARRL-CORR-JAN-2011.pdf

- *Interference Resolution | Federal Communications Commission* https://www.fcc.gov/enforcement/areas/interference-resolution#:~:text=Consumers%20or%20members%2 0of%20the,using%20the%20Consumer%20Complaint%20Center.

- *The Trouble with Practice Exams - HamTestOnline* https://www.hamradiolicenseexam.com/trouble-with-practice -exams.htm#:~:text=Practice%20exams%20do%20serve%20a,they%20should%20not%20replace%20education.